A posteriori error estimation for non-linear eigenvalue problems for differential operators of second order with focus on $3D$ vertex singularities

Von der Fakultät für Mathematik der Technischen Universität Chemnitz genehmigte

D i s s e r t a t i o n

zur Erlangung des akademischen Grades
Doctor rerum naturalium
(Dr. rer. nat.)

vorgelegt von Dipl.-Math. Cornelia Pester
geboren am 08.07.1978 in Karl-Marx-Stadt

eingereicht am 25. November 2005

Gutachter: Prof. Thomas Apel
Prof. Arnd Meyer
Prof. Serge Nicaise

verteidigt am 21. April 2006

Bibliografische Information Der Deutschen Bibliothek

Die Deutsche Bibliothek verzeichnet diese Publikation in der Deutschen Nationalbibliografie; detaillierte bibliografische Daten sind im Internet über http://dnb.ddb.de abrufbar.

ISBN 3-8325-1249-7

Logos Verlag Berlin
Comeniushof, Gubener Str. 47,
10243 Berlin
Tel.: +49 030 42 85 10 90
Fax: +49 030 42 85 10 92
INTERNET: http://www.logos-verlag.de

Preface

Boundary value problems of partial differential equations have a wide range of application in physics and engineering sciences, like heat conduction, metrology, elasticity, plasticity theory and thermodynamics. It is often necessary to evaluate the quality of certain materials and to predict their behavior under the influence of certain pressure, temperature and stresses. One interesting application is to forecast the onset and the propagation of cracks. Nowadays research work provides several methods for the evaluation and the prediction of the material behavior. For instance, the inner-material structure of a surface-breaking crack can be determined with certain ultrasonic measuring methods. Alternatively, such structures can be computed with mathematical models, which is cheaper and more efficient regarding, for example, varying material parameters. In the example of the surface-breaking crack, it is known that the angle between the crack line and the surface follows the typical $\sqrt{r}$-singularity. This means that the solution to the linear elasticity problem comprises terms of the form $r^{1/2}u(\underline{\mathbf{x}})$, where r is the distance from the crack tip at the surface. In order to determine the angle ξ of the interior crack line to the surface, the form $r^\alpha u$ is used for the solution to the elasticity problem in the mathematical model. The so-called singularity exponent α can be computed by numerical methods in dependence of the angle ξ and the material parameters. It is a kind of non-linear equation to determine those angles ξ for which the exponent α becomes $1/2$.

The mathematical model of this example belongs to the field of partial differential equations with vertex singularities. The theoretical background is embedded in the area of continuum mechanics which deals with the average behavior of elastic materials, known as elasticity theory. The linear elasticity problem for three-dimensional domains with concave corners is the major application of the theory developed in this thesis. The typical structure of the displacement function near the corner is used to derive an associated eigenvalue problem whose solutions quantify the singularities of the displacement function. This eigenvalue problem is defined on the unit sphere. For its discretization and the numerical solution, the finite element method is employed. In order to determine the quality of the approximate solution, the approximation error has to be estimated, because the exact eigenpairs are usually not known. This thesis is concerned with the theory of a posteriori error estimation for general eigenvalue problems on the sphere.

This work benefitted from the cooperation and the support by Thomas Apel who deserves special acknowledgment for many fruitful discussions and numerous useful ideas. The theory presented here involves details from various fields of mathematics and mechanics. I am particularly indebted to Arnd Meyer, Bernd Hofmann and Bernd Silbermann for their valuable advice. My friends and colleagues shall be gratefully acknowledged as well for their helpful remarks and useful suggestions; in particular, Dominique Leguillon, Anton Heinen, Gunter Winkler, Gerd Kunert, Tino Eibner, Sergey Grosman and Matthias Pester. Furthermore, I thank the DFG (German research foundation) for the financial support of this project under the grant numbers AP 72/1-1,2,3.

Keywords: a posteriori error estimation, Clément-type interpolation, corner singularities, unit sphere, nonlinear eigenvalue problems, spectral theory

AMS(MOS): 35J05; 58J50; 65N15; 65N25; 65N30

Contents

List of symbols

The list below gives an overview of the notation which is frequently used accompanied by a brief explanation and (where possible) a page number of its definition or major occurrence. In general, we write scalars (real or complex numbers) and scalar functions in italic type, vector functions in underlined italic type and special vectors (or points in $\mathbb{R}^3$) in boldface roman type. The index $\mathcal{S}$ is used to indicate that the vector or operator under consideration is defined on the unit sphere.

Furthermore, the notation $a \sim b$ and $a \lesssim b$ means that there are constants $c_1, c_2 > 0$ (which are independent of the discretization and the function under consideration), so that $c_1\, b \leq a \leq c_2\, b$ and $a \leq c_2 b$, respectively.

Symbol	Description	Page
$\mathcal{S}^2$	unit sphere in $\mathbb{R}^3$	
Ω	infinite conical domain in $\mathbb{R}^3$ with a polyhedral corner	5
$\partial\Omega_D$, $\partial\Omega_N$	Dirichlet and Neumann boundary of Ω	6
$\Omega_\mathcal{S}$	connected subset of the unit sphere, $\Omega_\mathcal{S} = \Omega \cap \mathcal{S}^2$	35,74
Γ_D, Γ_N	Dirichlet and Neumann boundary of $\Omega_\mathcal{S}$, $\Gamma_{D/N} = \partial\Omega_{D/N} \cap \mathcal{S}^2$	35
ξ_1, ξ_2	parametrization of $\Omega_\mathcal{S}$	30
φ, θ, r	spherical coordinates	38
$\tilde{\Omega}_\mathcal{S}$	the parameter domain corresponding to $\Omega_\mathcal{S}$	38
$\underline{\mathbf{X}}$	point in $\mathbb{R}^3$, $r := \lvert\underline{\mathbf{X}}\rvert$ = distance from the corner	30
$\underline{\mathbf{x}}$	point on the unit sphere $\underline{\mathbf{x}} = \underline{\mathbf{X}}/\lvert\underline{\mathbf{X}}\rvert$	30
h	global mesh size, discretization parameter	36
$\mathcal{T}_h$	(isotropic) triangulation of $\Omega_\mathcal{S}$	36,41
$\mathcal{E}_h$, $\mathcal{E}(T)$ etc.	sets of edges	36
$\mathcal{N}_h$, $\mathcal{N}(T)$ etc.	sets of nodes (or vertices)	36
$\mathcal{Z}$	maximum number of elements with one common vertex	42,59
x	node (or vertex) in the triangulation, $\mathrm{x} \in \mathcal{N}_h$	
$\omega_\mathrm{x}, \omega_E, \omega_T, \tilde{\omega}_E, \tilde{\omega}_T$	patches of elements	36
ϕ_x	(spherical) nodal basis functions	37,44
$\mathfrak{b}_T$, $\mathfrak{b}_E$	element and edge bubble functions	50
$\mathbf{n}$	normal vector to a two-dimensional manifold	32
$\mathbf{n}_\mathcal{S}$	normal vector to a curve on the sphere	32
$\mathbf{n}_{T,E}$	normal vector to an edge E (exterior to the element T)	73
$\mathbf{n}_E$	normal vector to an edge E (arbitrary direction)	74
$\lfloor\psi\rfloor_E$	jump of a function ψ over an edge E into the direction $\mathbf{n}_E$	74
$\mathrm{d}\Omega$, $\mathrm{d}\Sigma$	(Cartesian) volume and surface elements	31
$\mathrm{d}\mathcal{S}$, $\mathrm{d}\sigma$	(spherical) surface and line elements	31
$\nabla_\mathcal{S}$	spherical gradient	31

$\Delta_{\mathcal{S}}$	Laplace-Beltrami operator	77
$\mathcal{H}^k(\omega)$, $[\mathcal{H}^k(\omega)]^3$	(weighted) Sobolev spaces of functions on $\omega \subset \mathcal{S}^2$, $k = 0, 1$	32
$L^2(\omega)$	$L^2(\omega) = \mathcal{H}^0(\omega)$ if $\omega \subset \mathcal{S}^2$; $L^2(\omega) = H^0(\omega)$ otherwise	
V	space of $[\mathcal{H}^k(\Omega_{\mathcal{S}})]^d$-functions which vanish on Γ_D	35
V_h	finite element space	36
$\tilde{V}_h$	$\tilde{V}_h = \operatorname{span}\{\mathfrak{b}_T, \mathfrak{b}_E \mid T \in \mathcal{T}_h,\ E \in \mathcal{E}_h\}^d \subset V$	50
$\\|\lceil\cdot\rceil\\|_k$, $\lceil\cdot\rceil_k$	(weighted) norms and seminorms	31
$\\|\cdot\\|_k$, $\|\cdot\|_k$	standard Sobolev norms	31
$\lceil\omega\rceil$	spatial size of the domain $\omega \subset \mathcal{S}^2$, $\lceil\omega\rceil = \\|\lceil 1\rceil\\|^2_{0,\omega}$	32
$\lceil\gamma\rceil$	spatial length of the curve $\gamma \subset \mathcal{S}^2$, $\lceil\gamma\rceil = \\|\lceil 1\rceil\\|^2_{0,\gamma}$	32
$\mathbb{R}_+$	set of all positive real numbers	75,83
$\mathbb{K}$	$\mathbb{K} = \mathbb{R}$ or $\mathbb{K} = \mathbb{C}$	
$\mathcal{L}(X, Y)$	set of linear, bounded operators $\mathcal{A} : X \to Y$	12
$\ker \mathcal{A}$, $R(\mathcal{A})$	kernel and range of the operator $\mathcal{A}$	13
$\mathcal{A}^\#$, $\mathcal{A}^\star$	dual and adjoint operators corresponding to $\mathcal{A}$	13
$\mathcal{B}(\cdot) : \mathbb{K} \to \mathcal{L}(V,V)$	pencil of linear, bounded Fredholm operators	19
F, F_h, $\tilde{F}_h$	(nonlinear) functionals which are defined by the eigenvalue problem and its discretization	21,37,50
$\mathrm{D}F(x_0)$	(Fréchet) derivative of an operator F at x_0	15
$\pi_{0,\omega}$	L^2-projection on the sphere, $\pi_{0,\omega} v = (\int_\omega \mathrm{d}\mathcal{S})^{-1} (\int_\omega v \, \mathrm{d}\mathcal{S})$	45
I_h	Clément-type interpolation operator	45
λ	eigenvalue of the operator pencil $\mathcal{B}(\cdot)$ or of an eigenvalue problem with shifted spectrum, in general $\lambda = \alpha + \frac{1}{2}$	7,86
α	singularity exponent, in general $\alpha = \lambda - \frac{1}{2}$	7,75,82
u, $\underline{u}$	usually an eigenfunction of $\mathcal{B}(\cdot)$	7,20,35
v, $\underline{v}$	usually a test function in the variational formulation of the eigenvalue problem	21,36,83
$\vartheta_{+,T}$, $\vartheta_{-,T}$	$\vartheta_{+,T} = \sup_{\mathbf{x}(\varphi,\theta) \in T} \sin\theta$, $\vartheta_{-,T} = \inf_{\mathbf{x}(\varphi,\theta) \in T} \sin\theta$	41
$h^{\mathcal{S}}_{\varphi,T}$, $h^{\mathcal{S}}_{\theta,T}$	spatial dimensions of T, see Definition 3.4	41
$h_{+,T}$, $h_{-,T}$	$h_{+,\omega} = \max\{h^{\mathcal{S}}_{\varphi,\omega}, h^{\mathcal{S}}_{\theta,\omega}\}$, $h_{-,\omega} = \min\{h^{\mathcal{S}}_{\varphi,\omega}, h^{\mathcal{S}}_{\theta,\omega}\}$	41
δ_T	diameter of the spherical incircle of T, see Definition 3.4	41
$\underline{R}_T$, $\underline{R}_E$	element and edge residuals	46,78,92
$\underline{R}^\star_T$, $\underline{R}^\star_E$	residuals of the adjoint problem	58,92
$\tilde{\eta}_T, \eta_T$	error estimators for the primal eigenvalue problem	47,49
$\tilde{\eta}^\star_T, \eta^\star_T$	error estimators for the adjoint eigenvalue problem	57,58
$\varepsilon_T, \varepsilon^\star_T$	approximation errors for the primal and the adjoint problems	49,57
$\varepsilon_{\mathcal{S}}$, ε_3	special strain tensors on the sphere	83
$\sigma_{\mathcal{S}}$	special stress tensor on the sphere	83
A	material tensor (order 4) for the linear elasticity problem, real and piecewise constant	82,98
$C_{T,adj}$	constant in $\lceil T_1\rceil \le C_{T,adj} \lceil T_2\rceil$ for adjacent elements T_1, T_2	60
C_{h_+}	constant in $h_{+,T_1} \le C_{h_+} h_{+,T_2}$ for adjacent elements T_1, T_2	60
$C_{1,V}$, $C_{V,1}$	constant in $\lceil\underline{v}\rceil_{1,\Omega_{\mathcal{S}}} \le C_{1,V} \\|\underline{v}\\|_V$ and $\\|\underline{v}\\|_V \le C_{V,1} \\|\lceil\underline{v}\rceil\\|_{1,\Omega_{\mathcal{S}}}$	59
$C_{H,V}$	constant in $\\|\underline{v}\\|_H \le C_{H,V} \\|\underline{v}\\|_V$	20,91,94

1 Introduction

The theory of elliptic partial differential equations has its origin in the middle of the eighteenth century, when d'Alembert published an essay on the new theory of the resistance of fluids [27]. At about the same time, Euler derived the equations of motion for an irrotational fluid in three dimensions [37]. A brief overview of the historical development up to the beginning of the twentieth century is given by McLean [71] including the general ideas of the contributions by Lagrange, Laplace and Poisson.

Boundary value problems of partial differential equations have a wide range of application, for instance, in physics and engineering sciences including heat transfer and elasticity problems. The boundary element method [99, 108] and the finite element method [19, 20, 23] are powerful means to solve partial differential equations. Among the boundary value problems, the linear elasticity theory is of vast interest. It is impossible to give a satisfying overview of the literature published on this topic. Only a few books shall be mentioned. Particular attention shall be drawn to the book by Love [67] which starts with an interesting historical introduction to the elasticity theory. The first experiments by Galilei with a beam in bending, Hooke's peculiar name for his famous law, the introduction of Young's modulus, the ups and downs in the development of the elasticity theory until the pioneering work by Green and Cauchy and the conclusive investigations of Saint-Venant and Lord Kelvin are just a few stations in the worth reading historical background. A nice introduction to the elasticity and plasticity theory is given in the book by Parisch [85], including a comprehensible overview of tensor calculus and further notation, the solution with the finite element method and some applications. Earlier publications are, for example, [65, 77]. The book by Malvern [68] provides an extensive overview of mechanics of a continuous medium. The Handbook of Elasticity Solutions [47] contains a collection of elasticity solutions in topics of solid mechanics and quantitive material science, such as contact problems, Green's functions for isotropic and anisotropic solids or cracks in two- and three-dimensional solids.

Three-dimensional crack problems belong to the main applications of the scientific work around this thesis. In general, we deal with the solutions to elliptic boundary value problems near the vertex of a polyhedral cone, that is, in the neighborhood of a three-dimensional, concave corner, with special emphasis to a posteriori error estimation. Elliptic boundary value problems near corners were studied, for instance, in [8, 9, 14, 28, 41, 55]. The solutions to such problems are not necessarily smooth in the sense of the typical Sobolev regularity; it is known that stress singularities can arise near the vertex [28, 70, 79]. Analytic considerations concerning the structure and explicit formulae for the singularities were presented, for example, by Sändig et al. [14, 82, 97, 98], see also [42]. A comprehensive analysis of the regularity properties of these solutions was done, for example, in [28, 51, 59, 79], see also [55] and references therein. The singularities of the solutions are specified by so-called singularity exponents. The computation of these components was analyzed, for instance, in [13, 30, 31, 32, 33, 34, 100, 102]. Since singularities are of local nature, we can assume, without loss of generality, that the given domain is smooth except for one angular point, see Figure 1.1. Let $\Omega \subset \mathbb{R}^3$ be such a domain. The coordinate system shall be chosen so that the corner is placed at the origin $(0,0,0)$. In addition, we assume, that Ω coincides with a cone in a neighborhood of the corner. This means that there is a constant $R > 1$ so that Ω equals the set $\mathcal{K}_R := \{\underline{\mathbf{X}} \in \mathbb{R}^3 \mid \underline{\mathbf{X}}/|\underline{\mathbf{X}}| \in \Omega \cap \mathcal{S}^2,\, 0 < |\underline{\mathbf{X}}| < R\}$ in some neighborhood of the origin, i.e. $\mathcal{K}_R \cap \Omega = \mathcal{K}_R$, see Figure 1.1.

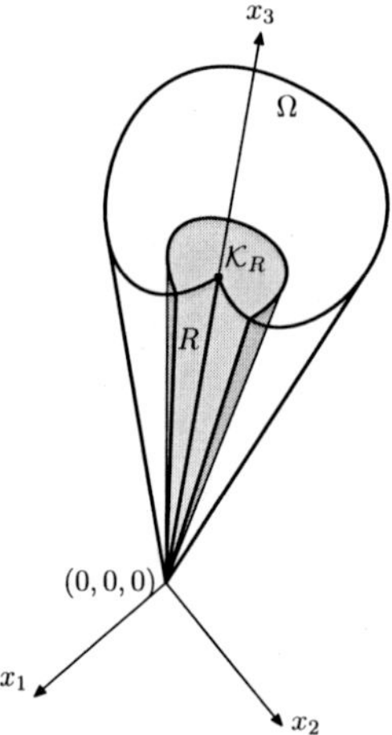

Figure 1.1: Conical domain: Ω coincides with a cone in the neighborhood of the vertex

When L is a differential operator and $f \in L^2(\Omega)$, let us consider the mixed boundary value problem

$$\begin{aligned} \mathrm{L}U &= f \quad \text{in } \Omega, \\ U &= 0 \quad \text{on } \partial\Omega_D, \\ \ell U &= g \quad \text{on } \partial\Omega_N, \end{aligned} \tag{1.1}$$

where $\partial\Omega_D$ and $\partial\Omega_N$ denote the Dirichlet and Neumann boundary parts of $\partial\Omega$, respectively, and where ℓ is an operator defining the Neumann boundary conditions. Let U be a solution to (1.1). We are interested only in the structure of U near the corner, so that we can assume further, without loss of generality, that $\Omega = \mathcal{K}_\infty$ is an *infinite conical domain.*

In general, the solution U to problem (1.1) does not possess the typical Sobolev regularity. If the differential operator L is of second order, then U is composed of a regular part $U_r \in H^2(\Omega)$ and a singular part $U_s \notin H^2(\Omega)$. If the right-hand side F in (1.1) is an L^2-function, but U_s has not the sufficient smoothness, then U_s must be mapped to zero by L. Likewise, $\ell U_s = 0$ on $\partial\Omega_N$. Thus, we consider the modified boundary value problem

$$\begin{aligned} \mathrm{L}U &= 0 \quad \text{in } \Omega, \\ U &= 0 \quad \text{on } \partial\Omega_D, \\ \ell U &= 0 \quad \text{on } \partial\Omega_N, \end{aligned} \tag{1.2}$$

where we assume that $\Omega = \mathcal{K}_\infty$ is an infinite conical domain.

In the pioneering work by Kondrat'ev [51], it was shown that the singular parts of the solutions to elliptic boundary value problems can be developed in each point $\underline{\mathbf{X}}$ near the corner into a series with terms of the form

$$c_i r^{\alpha_i} u_i(\alpha_i; \underline{\mathbf{x}})$$

and further logarithmic terms, where c_i are certain constants, $r = |\underline{\mathbf{X}}|$ is the distance from the corner and α_i are the so-called *singularity exponents* of the solution.

The function u depends on the singularity exponent α and on $\underline{\mathbf{x}} = \underline{\mathbf{X}}/|\underline{\mathbf{X}}|$. The notation will be clarified more detailed in Section 2.4. For the sake of compactness, we will only write $u(\underline{\mathbf{x}})$ and assume the dependence on α tacitly. The singularities of the solution are quantified best by the exponents with smallest positive real part. Consequently, the search for functions U of the form

$$U(\underline{\mathbf{X}}) = r^{\alpha} u(\underline{\mathbf{x}}) \tag{1.3}$$

which satisfy (1.2) will be the starting point of the further considerations.

It turns out that the exponents α and the functions u are eigenpairs of an operator pencil $\mathcal{B}_{\mathrm{L}}$ which is defined by the given problem and its boundary conditions. This means, one is interested in the solutions $[\alpha, u]$ to the problem

$$\mathcal{B}_{\mathrm{L}}(\alpha) u = 0. \tag{1.4}$$

Neither α nor u depend on the variable r, so that the eigenvalue problem (1.4) is defined on the unit sphere. When the differential operator L in (1.1) has the order $2m$, then the operator pencil $\mathcal{B}_{\mathrm{L}}(\cdot)$ is a polynomial in α of the same order. Depending on the boundary value problem (1.1), problem (1.4) is not necessarily symmetric. This means that there might be complex eigenvalues α and complex-valued eigenfunctions u. Because of the real structure of the original problem, the eigenvalues and eigenfunctions appear in conjugate complex pairs so that the complex addends in the series expansion of U_s are extinguished. The behavior of the factor r^{α} near the corner was studied, for instance, in [46]. We will concentrate on the solutions $[\alpha, u]$ to problem (1.4).

The spectrum of the eigenvalue problem (1.4) which is associated with the given boundary value problem in a conical domain has certain symmetry properties. It was stated in [55, Theorem 1.2.2], see also [54], that under certain conditions the eigenvalues appear in quadruplets which are symmetric with respect to the real axis and to some axis $\operatorname{Re} \alpha = \gamma/2$. In the case of three-dimensional corner singularities, we have $\gamma = -1$. From the numerical and the algebraic point of view, it would be preferable if $\gamma = 0$, this means that the spectrum is symmetric with respect to the real and the imaginary axes. Then, the discretized eigenvalue problem is equivalent to a standard matrix eigenvalue problem for a Hamiltonian or skew-Hamiltonian matrix, and special structure-preserving Lanczos or Arnoldi algorithms can be applied for an efficient computation of the singularities [3, 4, 15, 72, 73, 106, 105]. Moreover, a (skew-)Hamiltonian perturbation theory [52, 58, 103] ensures that the computation of the eigenvalues becomes more stable than with non-structured methods.

The desired spectral shift can be achieved by the substitution $\lambda = \alpha + \frac{1}{2}$ or by using the approach $r^{\lambda - 1/2} u$ instead of (1.3). Indeed, the resulting eigenvalue problem possesses the expected quadruplet structure, where λ is an eigenvalue if and only if $\bar{\lambda}$, $-\lambda$ and $-\bar{\lambda}$ are eigenvalues, too. The introduction of the parameter λ was first suggested in [7], where a priori error estimates for the finite element method on graded meshes were derived.

The mentioned eigenvalue problem is usually obtained from a Mellin transformation with respect to the variable r. Another possibility is presented for the linear elasticity problem in [64], where the approach (1.3) is inserted directly into the given boundary value problem. This alternative way is more straightforward and can be applied to other boundary value problems, as well, see [55, 75]. In either case, it is necessary to intersect the given domain with a ball which is centered at the corner. Since neither α nor u depend on the radial variable (the distance from the corner), the eigenvalue problem for α and u is defined on the unit sphere $\mathcal{S}^2$. For the finite element discretization of this eigenvalue problem, it is necessary to parametrize the sphere. It lies in the nature of a spherical domain to be parametrized with spherical coordinates as it is often done in the literature on vertex singularities, see [7, 55, 64, 14, 98]. From the numerical point of view, spherical coordinates are not necessarily the best choice, because they produce artificial poles. The frequently suggested alternatives, however, yield difficulties as well or are more restrictive. For example, the projection of triangles in $\mathbb{R}^3$ with nodes on $\mathcal{S}^2$ onto the sphere is applicable only to spherical domains with geodesic boundary, so that simple domains, like the intersection of a circular cone with the unit ball, are excluded immediately. The discussion of the pros and cons of further parametrizations is postponed to Section 3.2. For the proof of certain estimates for the discretized eigenvalue problem, it is necessary to settle some assumptions on the finite element mesh and therefore to fix a parametrization. In spite of the hurdles near the poles, we will use spherical coordinates for this. Nevertheless, we will keep the notation in this work rather general so that no parametrization is excluded entirely.

Numerical methods for the computation of the solutions to boundary value problems in non-smooth domains or on arbitrary two-dimensional surfaces were described, for example, in [8, 36, 111]. Our aim is to estimate the error between the solutions to a given eigenvalue problem on the sphere and the solutions to the corresponding discretized eigenvalue problem. For Cartesian domains, the literature on eigenvalue and boundary value problems and appropriate strategies for the a priori and a posteriori error estimation is extensive, including [1, 10, 44, 62, 80, 104]. We will derive a reliable and efficient residual a posteriori error estimator for the eigenpairs $[\alpha, u]$ of (1.4), where we adapt the strategy proposed in the monograph by Verfürth [104]. The theory in the present work is applicable not only to eigenvalue problems on the sphere, but also includes results for usual two-dimensional domains and for two-dimensional manifolds, provided that the given eigenvalue problem satisfies certain analytic properties and that an admissible triangulation of the domain under consideration is used for the finite element discretization. Details concerning these assumptions are given in Sections 2.3, 3.1 and 3.2. Since most results are well-known for the planar, two-dimensional domains, the focus of this thesis is put on spherical domains. Although the route is very similar to the theory for the planar case, special care has to be taken in the verification of otherwise standard theorems (e.g. trace theorem, Poincaré inequality and interpolation error estimates), because the parametrization of the sphere and the corresponding spherical surface and line elements require the introduction of weighted norms and operators. The appropriate weights depend on the metric fundamental terms of the used parametrization.

Moreover, the presented theory is not restricted to linear eigenvalue problems. On the contrary, we obtain results for arbitrary operator pencils. Having in mind the Laplace and the linear elasticity problems as special applications to our theory, we are particularly interested in such operator pencils which are quadratic or linear with respect to the eigenvalue.

Another milestone of this thesis is the consideration of *complex*-valued *vector* functions, which occur, for instance, in the analysis of the linear elasticity problem near the vertex of a cone. It will turn out that the theory for eigenvalue problems with real-valued solutions has to be modified in several details so that all results have to be proven anew. Some of these changes are discussed in Section 2.3.

The numerical analysis of eigenvalue problems with complex solutions requires details from various fields of mathematics beside numerics, including functional and complex analysis, Fredholm and function theory as well as differential geometry and tensor calculus. In order to present a self-contained work, Chapter 2 is devoted to an introduction to operator theory and differential geometry, where we concentrate on the terminology and properties which are essential in the forth-coming analysis of the eigenvalue problems. Nevertheless, Chapter 2 is not only a repetition of known facts. We enrich the basic theory with conclusions which will be used later in the verification of a posteriori error estimates. In Section 2.2, we present already first approximation results for the solutions to non-linear operator problems; and in Section 2.3, we prove an important relation between the multiplicity of an eigenvalue and the invertibility of the Fréchet derivative of the operator which is associated with the given eigenvalue problem.

Chapter 3 is concerned with error estimation. In [7], Apel, Sändig and Solov'ev derived a priori error estimates for $3D$ vertex singularities for the linear elasticity problem based on a finite element method on graded meshes. In this work, we derive a residual a posteriori error estimator. It turns out that the meshes suggested in [7] are not suited for a posteriori error estimation, although sufficient for a priori estimates. This shortcoming is caused by an overrefinement near the poles of the ball which results in a reliability–efficiency gap of the a posteriori error estimator. We will therefore use a new triangulation method of the given domain, which will be described in detail in Section 3.2. Numerical experiments show that one can even do without the graded meshes, when the mesh is refined adaptively, because the error is largest near the angular points of the domain, so that an adaptive refinement strategy will refine the mesh near these points anyway.

Essential ingredients in the error estimation are interpolation error estimates. In Section 3.3, we introduce a (weighted) Clément-type interpolation operator [24] which interpolates constant functions exactly.

Beside further auxiliary results (trace theorem, Poincaré-type and Friedrichs-type inequality on spherical domains), we formulate the local interpolation error estimates

$$\begin{aligned} \lVert\!\lceil \underline{v} - \mathrm{I}_h \underline{v} \rceil\!\rVert_{0,T} &\leq C_{\mathrm{I}_h,T}\, h_{+,T} \lceil \underline{v} \rceil_{1,\tilde{\omega}_T} \quad \forall \underline{v} \in [\mathcal{H}^1(\tilde{\omega}_T)]^d, \\ \lVert\!\lceil \underline{v} - \mathrm{I}_h \underline{v} \rceil\!\rVert_{0,E} &\leq C_{\mathrm{I}_h,E} \frac{\lceil E \rceil^{1/2}}{\lceil T \rceil^{1/2}} h_{+,T} \lceil \underline{v} \rceil_{1,\tilde{\omega}_E} \quad \forall \underline{v} \in [\mathcal{H}^1(\tilde{\omega}_E)]^d. \end{aligned}$$

in Theorem 3.11 which we prove together with the auxiliary results in Section 3.8. This is the only part where the parametrization of the sphere with spherical coordinates is exploited explicitly. The symbols will be explained in Sections 2.4, 3.1 and 3.2.

Finally, we derive a residual a posteriori error estimator for the error $|\lambda_0 - \lambda_h| + \|\underline{u}_0 - \underline{u}_h\|_V$, where $[\lambda_h, \underline{u}_h]$ is a finite element approximation to the exact eigenpair $[\lambda_0, \underline{u}_0]$, and prove its reliability and efficiency in Sections 3.4 and 3.5. A priori error estimates reveal that the eigenvalues converge faster than the eigenfunctions [7]. In Section 3.6, a separate estimate for the eigenvalues is derived, based on additional approximation results which were formulated, for instance, in [49]. To this end, it is necessary to consider the adjoint eigenvalue problem. Due to certain symmetry properties of the eigenvalue problems which are associated with the Laplace and the linear elasticity problems, there is a strong relation between the primal and the adjoint discretized eigenvalue problems. We show that

$$\|\underline{u}_0 - \underline{u}_h\|_V \leq c\,\eta, \qquad \|\underline{u}_0^\star - \underline{u}_h^\star\|_V \leq c\,\eta^\star, \qquad |\lambda_0 - \lambda_h| \leq c\,\eta\eta^\star,$$

where η and $\eta^\star$ are the residual error estimators for the primal and the adjoint problem, respectively, and where $[\lambda_h, \underline{u}_h^\star]$ is a finite element approximation of the eigenpair $[\lambda_0, \underline{u}_0^\star]$ of the adjoint eigenvalue problem. The constant c is generic. While the computed a posteriori error estimator η is reliable and efficient, we obtain merely an upper bound for the eigenvalues.

During the derivation of the error estimates for eigenvalue problems, the constants in the appropriate inequalities were traced. A strategy to find the constants in the interpolation error estimates and upper error bounds for boundary value problems on planar, two-dimensional domains was suggested in [22]. The proposed method, however, does not work for complex vector functions and is restricted to special two-dimensional domains. In the present work, other techniques were used, mainly based on the consideration of reference domains. Moreover, it is also possible to give estimates for the constants in the lower error bound. Owing to the spherical nature of the problems under consideration, the derivation of the constants is technical and complicated to such an extent that it would go beyond the scope of this thesis to present all details. This is why only the main results of this research are presented in Section 3.7 without proof.

In Chapter 4, we discuss two model problems, the Laplace and the linear elasticity problem. The results are justified with numerical experiments preceded by an explanation how the implementation of the residual error estimator in spherical coordinates was realized.

The notation used throughout this thesis is summarized on the pages 3–5.

2 Preliminaries

2.1 Complex analysis and operator theory

In this section, we repeat definitions and facts from functional analysis, operator theory, complex analysis and function theory which will be essential in the following numerical analysis of eigenvalue problems. While most things should be known for vector spaces over the real numbers, things become more restrictive in the complex case. In the formulation of complex theory, one usually considers the so-called complexification of a real vector space and generalizes the results of the real case to complex vector spaces. We will do the same and refer to basic literature on functional and complex analysis, including [2, 26, 61, 90, 95, 109, 112, 113], for more detailed information. A nice overview with numerical application is also given in [71].

Sesquilinear forms play an important role in the whole work. We remark that it is common in physical applications to define sesquilinear forms so that they are linear with respect to the second argument and antilinear (or conjugate linear) with respect to the first argument [43, 92, 107]. In mathematical applications, however, the opposite convention is standard. In the following, we will use sesquilinear forms which are linear with respect to the first argument and antilinear with respect to the second argument. The same convention holds for inner products which are special sesquilinear forms.

Real and complex Hilbert spaces. Let X be a vector space over $\mathbb{R}$. We say that X is a *real Hilbert space* or a *Hilbert space over* $\mathbb{R}$, if it is equipped with an inner product $(\cdot,\cdot)_X : X \times X \to \mathbb{R}$ and if X is complete with respect to the norm $\|x\|_X = \sqrt{(x,x)_X}$.

Suppose that X_R is a real Hilbert space with the inner product $(\cdot,\cdot)_{X_R} : X_R \times X_R \to \mathbb{R}$. The space $X := X_R \times X_R = \{[x_R, x_I] \mid x_R, x_I \in X_R\}$ with the operations

$$\begin{aligned} [x_R, x_I] + [y_R, y_I] &= [x_R + y_R, x_I + y_I], && x_R, x_I, y_R, y_I \in X_R, \\ (\alpha + \mathrm{i}\beta)[x_R, x_I] &= [\alpha x_R - \beta x_I, \beta x_R + \alpha x_I], && \alpha, \beta \in \mathbb{R},\ x_R, x_I \in X_R, \end{aligned}$$

is called the *complexification* of X_R, cf. [96]. Each element $[x_R, x_I] \in X_R \times X_R$ can be represented by the sum $[x_R, 0] + \mathrm{i}[x_I, 0]$. Identifying $[x_R, 0] \in X$ with $x_R \in X_R$, we have formally that $X_R \subset X_R \times X_R$ and can write $[x_R, x_I] = x_R + \mathrm{i}x_I$. The space $X = X_R \times X_R$ is a vector space over $\mathbb{C}$ and equipped with the inner product $(\cdot,\cdot)_X : X \times X \to \mathbb{C}$,

$$([x_R, x_I], [y_R, y_I])_X := (x_R, y_R)_{X_R} + (x_I, y_I)_{X_R} + \mathrm{i}\Big((x_I, y_R)_{X_R} - (x_R, y_I)_{X_R}\Big). \tag{2.1}$$

We call such a space X a *complex Hilbert space* or a *complexified Hilbert space* or a *Hilbert space over* $\mathbb{C}$.

Let $x = [x_R, x_I] \in X = X_R \times X_R$. The conjugate complex element will be denoted by $\bar{x} = [x_R, -x_I]$. If $x \in X_R$ (i.e. $x_I = 0$), then $\bar{x} = x$.

Lemma 2.1. *Let $X = X_R \times X_R$ be a complexified Hilbert space. The form $(\cdot,\cdot)_X$ satisfies the inner product properties,*

$$\begin{array}{lcl} (x,y)_X &=& \overline{(y,x)_X} \qquad \forall x,y \in X \\ (\gamma x,y)_X &=& \gamma(x,y)_X \qquad \forall \gamma \in \mathbb{C},\ \forall x,y \in X \\ (x,x)_X = 0 &\iff& x = 0, \quad (x,x)_X > 0 \quad \forall x \in X \setminus \{0\}. \end{array}$$

Moreover, the definition of $(\cdot,\cdot)_X$ implies that

$$\overline{(x,y)_X} = (\bar{x},\bar{y})_X. \tag{2.2}$$

Proof. Note that $(x_R,y_R)_{X_R} = (y_R,x_R)_{X_R}$ since $(x_R,y_R)_{X_R} \in \mathbb{R}$ for all $x_R \in X_R$. Let $x,y \in X$ be arbitrary. There are element $x_R, x_I, y_R, y_I \in X_R$ such that $x = [x_R,x_I]$ and $y = [y_R,y_I]$. The first two relations are obtained by direct computation making use of the definition (2.1). Moreover, $(x,x)_X = 0$ if and only if $\mathrm{Re}(x,x)_X = 0$ and $\mathrm{Im}(x,x)_X = 0$, that is,

$$(x_R,x_R)_{X_R} + (x_I,x_I)_{X_R} = 0 \quad \text{and} \quad (x_I,x_R)_{X_R} - (x_R,x_I)_{X_R} = 0.$$

The latter condition is always true and thus $(x,x)_X \in \mathbb{R}$. From the inner product properties of $(\cdot,\cdot)_{X_R}$, we conclude that $(x_R,x_R)_{X_R} \geq 0$ and $(x_I,x_I)_{X_R} \geq 0$, and therefore, $(x_R,x_R)_{X_R} + (x_I,x_I)_{X_R} = 0$ if and only if $x_R = x_I = 0$, i.e., $x = [0,0] = 0$.

Finally, the relation (2.2) follows from

$$\begin{array}{rcl} \overline{(x,y)_X} &=& \overline{([x_R,x_I],[y_R,y_I])_X} \\ &=& (x_R,y_R)_{X_R} + (x_I,y_I)_{X_R} - \mathrm{i}\Big((x_I,y_R)_{X_R} - (x_R,y_I)_{X_R}\Big) \\ &=& (x_R,y_R)_{X_R} + (-x_I,-y_I)_{X_R} + \mathrm{i}\Big((-x_I,y_R)_{X_R} - (x_R,-y_I)_{X_R}\Big) \\ &=& ([x_R,-x_I],[y_R,-y_I])_X \;=\; (\bar{x},\bar{y})_X. \end{array}$$

□

Linear and dual operators. Let $\mathbb{K} = \mathbb{R}$ or $\mathbb{K} = \mathbb{C}$ and let X and Y be two Hilbert spaces over $\mathbb{K}$. An operator $\mathcal{A} : X \to Y$ is called a *linear operator* if $\mathcal{A}(x_1 + x_2) = \mathcal{A}x_1 + \mathcal{A}x_2$ for all $x_1, x_2 \in X$ and $\mathcal{A}(\gamma x) = \gamma \mathcal{A}x$ for all $x \in X$, $\gamma \in \mathbb{K}$. A linear operator is *bounded*, if $\|\mathcal{A}x\|_Y \leq M\|x\|_X$ for all $x \in X$ and some positive constant M.

The set of all linear, bounded operators $\mathcal{A} : X \to Y$ is denoted by $\mathcal{L}(X,Y)$. The infimum over all constants in the boundedness condition equals the operator norm

$$\|\mathcal{A}\|_{\mathcal{L}(X,Y)} = \sup_{x \in X\setminus\{0\}} \frac{\|\mathcal{A}x\|_Y}{\|x\|_X}.$$

A linear operator is bounded if and only if it is continuous [109, Satz II.1.2], [2, Lemma 3.1].

A *linear functional* is a special linear operator $f : X \to \mathbb{K}$. The application of a linear functional f to an element x in X is usually denoted by the *duality product* $\langle\cdot,\cdot\rangle$, that is,

$$f(x) = \langle f, x\rangle \qquad \forall x \in X.$$

Note that, contrary to the inner product on $X \times X$, the duality product is *linear* in both components. To emphasize the linearity in x, one sometimes writes $f(x) = \langle x, f\rangle$ instead of

$f(x) = \langle f, x \rangle$. We will use either the notation $f(x)$ or the notation $\langle f, x \rangle$ in accordance with [104] on which our numerical analysis is based.

For two linear functionals $f, g : X \to \mathbb{K}$ and a scalar number $\gamma \in \mathbb{K}$, one defines new linear functionals $f + g$ and γf via

$$\begin{aligned} \langle f + g, x \rangle &:= \langle f, x \rangle + \langle g, x \rangle, \\ \langle \gamma f, x \rangle &:= \gamma \langle f, x \rangle. \end{aligned}$$

With these operations, the set of all linear functionals on X is a vector space over $\mathbb{K}$, the so-called *dual space* of X. It will be denoted by $X^\star := \mathcal{L}(X, \mathbb{K})$.

For each element $x \in X$, the relation

$$\langle f, \bar{v} \rangle = (x, v)_X \qquad \forall v \in X \tag{2.3}$$

defines a functional $f \in X^\star$ which is linear with respect to v. We emphasize that the argument v has to be conjugated on the left-hand side to guarantee the linearity of f. Alternatively, one often associates the functional $\langle \tilde{f}, v \rangle = (v, x)_X = (\bar{x}, \bar{v})_X$ with x, see [2, 113]. In connection with sesquilinear forms, the version (2.3) is sometimes more convenient and was practiced, for instance, in [66].

Vice versa, the *Riesz representation theorem* [2, 112] states that for each linear, bounded functional $f \in X^\star$ there is a unique element $\tilde{x}_f \in X$ such that $\langle f, v \rangle = (v, \tilde{x}_f)_X$ for all $v \in X$. Due to (2.2), the element $x_f = \bar{\tilde{x}}_f$ satisfies $\langle f, \bar{v} \rangle = (\bar{v}, \bar{x}_f)_X = (x_f, v)_X$, which is in compliance with the definition of the linear functional in (2.3). This means that there is a conjugate linear isomorphism between the spaces X and $X^\star$, cf. [2].

For $\mathcal{A} \in \mathcal{L}(X, Y)$ and $f \in Y^\star$, let $g \in X^\star$ be given by $\langle g, x \rangle = \langle f, \mathcal{A}x \rangle$ for all $x \in X$. The operator $\mathcal{A}^\# : Y^\star \to X^\star$, $\mathcal{A}^\# f = g$ is called the *dual operator* corresponding to $\mathcal{A}$. It is given via the relation

$$\langle f, \mathcal{A}x \rangle = \langle \mathcal{A}^\# f, x \rangle \qquad \forall x \in X \;\; \forall f \in Y^\star.$$

One can show that $\mathcal{A}^\# \in \mathcal{L}(Y^\star, X^\star)$ and that $\|\mathcal{A}\|_{\mathcal{L}(X,Y)} = \|\mathcal{A}^\#\|_{\mathcal{L}(Y^\star, X^\star)}$, see [2].

If $X = Y$, let $(\cdot, \cdot)_X : X \times X \to \mathbb{K}$ be the inner product in X. For a linear, bounded operator $\mathcal{A} \in \mathcal{L}(X, X)$, the *adjoint operator* $\mathcal{A}^\star \in \mathcal{L}(X, X)$ is defined by the equality

$$(\mathcal{A}x, y)_X = (x, \mathcal{A}^\star y)_X \qquad \forall x, y \in X.$$

Remark 2.2. *In the book by McLean [71], the operator $\mathcal{A}^\#$ is called the transposed operator. This terminology, however, appears to be a bit unfortunate, since transposed operators are usually employed in the context of inner products (in particular, $X = Y$) or matrix operations.*

The dual and the adjoint operators are related via the Riesz representation theorem: For each linear, bounded $f \in X^\star$, there is a unique element $x \in X$ such that $(x, y)_X = \langle f, \bar{y} \rangle$ for all $y \in X$. The operator $\mathcal{J} : X \to X^\star$, $x \mapsto f = \mathcal{J}\bar{x}$, is a so-called isometric conjugate linear isomorphism [2]. Hence, $\mathcal{A}^\# = \mathcal{J}\mathcal{A}^\star\mathcal{J}^{-1}$ or $\mathcal{A}^\star = \mathcal{J}^{-1}\mathcal{A}^\#\mathcal{J}$ because of the identity

$$\langle \mathcal{A}^\# \mathcal{J}\bar{x}, \bar{y} \rangle = \langle \mathcal{J}\bar{x}, \mathcal{A}\bar{y} \rangle = (x, \overline{\mathcal{A}\bar{y}})_X = (\mathcal{A}\bar{y}, \bar{x})_X = (\bar{y}, \mathcal{A}^\star\bar{x})_X = (\overline{\mathcal{A}^\star\bar{x}}, y)_X = \langle \mathcal{J}\mathcal{A}^\star\bar{x}, \bar{y} \rangle.$$

The terms adjoint *and* dual *operator are therefore often used synonymously.*

We denote the *kernel* and the *range* of an operator $\mathcal{A} : X \to Y$ by $\ker(\mathcal{A}) = \{x \in X \mid \mathcal{A}x = 0\}$ and $R(\mathcal{A}) = \{y \in Y \mid \exists x \in X\colon \mathcal{A}x = y\}$, respectively. For $\mathcal{A} \in \mathcal{L}(X, X)$, it is a known result from functional analysis that the relations

$$\ker(\mathcal{A}) \oplus \overline{R(\mathcal{A}^\star)} = X \qquad \text{and} \qquad \ker(\mathcal{A}^\star) \oplus \overline{R(\mathcal{A})} = X \tag{2.4}$$

hold, see, for instance, [109, Satz III.4.5]. The unitarity (complex orthogonality) of $\ker(\mathcal{A})$ and $R(\mathcal{A}^\star)$ (or likewise $\ker(\mathcal{A}^\star)$ and $R(\mathcal{A})$) is easily verified: suppose that $x \in \ker(\mathcal{A})$ and $y \in R(\mathcal{A}^\star)$; then there is a $y^\star \in X$ with $y = \mathcal{A}^\star y^\star$ and

$$0 = (\mathcal{A}x, y^\star)_X = (x, \mathcal{A}^\star y^\star)_X = (x, y)_X.$$

Invertibility: injective, surjective and bijective operators. Let X and Y be two (real or complex) Hilbert spaces and let $\mathcal{A} \in \mathcal{L}(X,Y)$ be a linear, bounded operator. We say that $\mathcal{A}$ is *injective*, if for each $y \in Y$ there is at most one element $x \in X$ such that $\mathcal{A}x = y$. The operator $\mathcal{A}$ is called *surjective*, if for each $y \in Y$ there is at least one element $x \in X$ such that $\mathcal{A}x = y$. An operator which is both injective and surjective, is called *bijective*. If $\mathcal{A}$ is bijective, then for each $y \in Y$ there is a unique element $x \in X$ such that $\mathcal{A}x = y$. The *inverse operator* which assigns $x \in X$ to $y \in Y$ will be denoted by $\mathcal{A}^{-1}$, $x = \mathcal{A}^{-1}y$.

We say that the operator $\mathcal{A}$ is *invertible* or a *linear homeomorphism*, if it is bijective and if the inverse operator is bounded. That is $\mathcal{A}^{-1} \in \mathcal{L}(Y,X)$.

In some books, an operator $\mathcal{A} \in \mathcal{L}(X,Y)$ is called invertible, if for each $y \in R(\mathcal{A})$ there is a unique element $x \in X$ so that $\mathcal{A}x = y$. In this case, the invertibility is equivalent to the injectivity. The definition given here requires $R(\mathcal{A}) = Y$ and $\mathcal{A}^{-1} \in \mathcal{L}(Y,X)$ in addition.

Compact operators. Let X be a (real or complex) Hilbert space. An operator $\mathcal{A} \in \mathcal{L}(X,X)$ is called *compact*, if for each bounded sequence $\{x_n\}$ in X, there is a subsequence $\{x_{n_k}\}$ so that $\{\mathcal{A}x_{n_k}\}$ converges in X. In formulae, this means that $\mathcal{A}$ is compact, if

$$\|x_n\|_X < M < \infty \quad \forall n \quad \Longrightarrow \quad \exists x_{n_k}, y^\star \in X: \quad \|\mathcal{A}x_{n_k} - y^\star\|_X \to 0 \quad \text{as } k \to \infty.$$

The set of compact operators is a subspace of $\mathcal{L}(X,X)$. In particular, each scalar multiple of a compact operator or the sum of two compact operators results in another compact operator, see, for instance, [109, Satz II.3.2 (a)].

If $\mathcal{C} \in \mathcal{L}(X,X)$ is a compact operator and $\mathcal{A} \in \mathcal{L}(X,X)$ is a linear, bounded operator, then the superposition $\mathcal{C}\mathcal{A}$ is a compact operator, too. To verify this, we employ the definition of a compact operator and omit the index k in the subsequence x_{n_k}. Let $\{x_n\}$ be a bounded sequence in X. Then, $\{\mathcal{A}x_n\}$ is a bounded sequence as well, since $\|\mathcal{A}x_n\|_X \le \|\mathcal{A}\|_{\mathcal{L}(X,X)}\|x_n\|_X$ for each linear, bounded operator $\mathcal{A}$. Hence, there is a $y^\star \in X$ so that $\|\mathcal{C}\mathcal{A}x_n - y^\star\|_X \to 0$.

Fredholm operators. Let X be a (real or complex) Hilbert space. A linear, bounded operator $\mathcal{A} \in \mathcal{L}(X,X)$ with $R(\mathcal{A}) = \overline{R(\mathcal{A})}$ is called a *Fredholm operator* if $\dim\ker\mathcal{A} < \infty$ and $\dim\ker\mathcal{A}^\star < \infty$. The number $\operatorname{Ind}\mathcal{A} = \dim\ker\mathcal{A} - \dim\ker\mathcal{A}^\star < \infty$ is called the *index* of the Fredholm operator $\mathcal{A}$.

If $\mathcal{A} \in \mathcal{L}(X,X)$ is a Fredholm operator and $\mathcal{J} \in \mathcal{L}(X,X)$ is invertible, then $\mathcal{A}\mathcal{J}$ is a Fredholm operator, too. This can be proven as follows: $\mathcal{A}x = 0$ if and only if $\mathcal{A}\mathcal{J}y = 0$ for the unique element $y = \mathcal{J}^{-1}x \in X$ (which always exists because $\mathcal{J}$ is invertible). Moreover, when $X_x = \{zx \mid z \in \mathbb{K}\}$ is a subspace of $\ker\mathcal{A}$, then $J^{-1}X_x = \{zy = \mathcal{J}^{-1}(zx) \mid z \in \mathbb{K}\}$ is a subspace of $\ker(\mathcal{A}\mathcal{J})$, since $\mathcal{J}$ is a linear operator. Hence, $\dim\ker\mathcal{A} = \dim\ker(\mathcal{A}\mathcal{J})$. Furthermore, $\mathcal{A}^\star x = 0$ if and only if $(z, \mathcal{A}^\star x)_X = (\mathcal{A}z, x)_X = 0$ for all $z \in X$. With $z = \mathcal{J}y \in X$, this is equivalent to $(\mathcal{A}\mathcal{J}y, x)_X = (y, (\mathcal{A}\mathcal{J})^\star x)_X = 0$ for all $y \in X$, i.e. $(\mathcal{A}\mathcal{J})^\star x = 0$. Thus, $\ker\mathcal{A}^\star = \ker(\mathcal{A}\mathcal{J})^\star$ and therefore $\dim\ker\mathcal{A}^\star = \dim\ker(\mathcal{A}\mathcal{J})^\star$.

Lemma 2.3 ([90]). *For each linear, bounded operator $\mathcal{A} \in \mathcal{L}(X, X)$ the following properties are equivalent:*

(i) *$\mathcal{A}$ is a Fredholm operator with the index* $\operatorname{Ind} \mathcal{A} = 0$.

(ii) *There is a compact operator $\mathcal{C}$ and an invertible operator $\mathcal{J}$, so that $\mathcal{A} = \mathcal{J} - \mathcal{C}$.*

Usually, the Fredholm theory [2, 93, 96] is concerned with operators of the form $\mathcal{I} - \mathcal{C}$ only, where $\mathcal{I} \in \mathcal{L}(X, X)$ is the identity operator on X and $\mathcal{C} \in \mathcal{L}(X, X)$ is a compact operator. For an invertible operator $\mathcal{J}$ and a compact operator $\mathcal{C}$, the operator $\mathcal{J} - \mathcal{C} = (\mathcal{I} - \mathcal{C}\mathcal{J}^{-1})\mathcal{J}$ is a Fredholm operator as well.

Differentiable functions. The concept of differentiability of real-valued functions is assumed to be known to the reader. For complex-valued functions $f : \mathbb{C} \to \mathbb{C}$, this concept is more restrictive. Complex analysis and function theory [21, 61, 110] teach that a complex-valued function $f(z) = u(z) + \mathrm{i}v(z)$ is differentiable with respect to the complex variable $z = x + \mathrm{i}y$ if and only if the Cauchy-Riemann equations

$$\frac{\partial u}{\partial x} = \frac{\partial v}{\partial y}, \qquad \frac{\partial u}{\partial y} = -\frac{\partial v}{\partial x}. \tag{2.5}$$

are satisfied. The expressions *holomorphic* or *complex-analytic* functions are often used interchangeably with *differentiable* functions in $\mathbb{C}$ although analytic functions have several other meanings. A function which is differentiable at every point $z_0 \in \mathbb{C}$ is called holomorphic on $\mathbb{C}$. The Cauchy-Riemann equations provide a necessary and sufficient condition for a complex-valued function to be differentiable. A holomorphic function is differentiable infinitely often. Examples of non-holomorphic functions are $f(z) = \bar{z}$ and $f(z) = |z|^2$. The derivative of a scalar function $f : \mathbb{C} \to \mathbb{C}$ with respect to the complex variable z is defined as for real functions,

$$f'(z_0) = \lim_{z \to z_0} \frac{f(z) - f(z_0)}{z - z_0} = \lim_{h \to 0} \frac{f(z_0 + h) - f(z_0)}{h}.$$

Depending on the way h tends towards zero (e.g. along the real or imaginary axis), one obtains that $f'(z) = \frac{\partial u}{\partial x} + \mathrm{i}\frac{\partial v}{\partial x}$ or $f'(z) = \frac{\partial v}{\partial y} - \mathrm{i}\frac{\partial u}{\partial y}$. Thus, the limit exists only if the Cauchy-Riemann equations (2.5) are satisfied. The equations (2.5) imply that u and v must be *harmonic* functions, that is, $\Delta u = \Delta v = 0$.

The concept of differentiability can be generalized to functions in (real or complex) Hilbert spaces and also to operators of such functions. Speaking about derivatives in a Hilbert space, one usually refers to the *Fréchet derivative* or the *Gâteaux derivative.* If the Fréchet derivative exists, then so does the Gâteaux derivative and both are identical. In general, for two Hilbert spaces X and Y, the (Fréchet) derivative of a (linear or non-linear) operator $F : X \to Y$ at a point $x_0 \in X$ is defined as the linear, bounded operator $A \in \mathcal{L}(X, Y)$ which satisfies

$$\lim_{x \to x_0} \frac{\|F(x) - F(x_0) - A(x - x_0)\|_Y}{\|x - x_0\|_X} = 0$$

if it exists. Usually, the derivative at x_0 is denoted by $\mathrm{D}F(x_0) := A$ or $F'(x_0) := A$. Let $x_0 \in X$ be fixed. When the derivative of F at x_0 exists, it can be written as

$$\mathrm{D}F(x_0)(x) = \lim_{t \to 0} \frac{1}{t}\Big(F(x_0 + tx) - F(x_0)\Big)$$

for each $x \in X$. In the case of a real Hilbert space, t tends towards zero along the real axis. In a complex Hilbert space, all complex directions towards the origin are allowed, which is more restrictive. For instance, the derivative of the operator $F : x \mapsto \|x\|_X^2$ exists in a real Hilbert space, but not in a complex one.

2.2 First approximation results

In this section, we summarize some general approximation results which we will employ later to prove an upper and a lower bound for the approximation error of the solution to a given eigenvalue problem.

Let X and Y be two (real or complex) Hilbert spaces, and let $F : D \subset X \to Y^\star$ be a given operator which is differentiable on a convex subset D_0 of D. (Like in real vector spaces, a set D_0 is called convex, if for any two elements $x, y \in D_0$ the elements $x + t(y-x)$, $t \in [0,1] \subset \mathbb{R}$, belong to D_0 as well, cf. [61].) If $F(x_0) = 0$ for a fixed element $x_0 \in X$, then x_0 is called a *solution* to

$$F(x) = 0. \tag{2.6}$$

We call x_0 a *regular solution* to (2.6), if the Fréchet derivative $\mathrm{D}F(x_0)$ exists and is a linear homeomorphism, that is, $\mathrm{D}F(x_0) \in \mathcal{L}(X, Y^\star)$, $\mathrm{D}F(x_0)^{-1} \in \mathcal{L}(Y^\star, X)$.

Lemma 2.4 ([104, Proposition 2.1]). *Let $x_0 \in D \subset X$ be a regular solution to problem (2.6). We assume that $\mathrm{D}F : D \to \mathcal{L}(X, Y^\star)$ is Lipschitz continuous at x_0 with a constant $\gamma > 0$. This means that there is a number $R_0 > 0$ such that*

$$\gamma := \sup_{x \in B_{R_0}(x_0)} \frac{\|\mathrm{D}F(x) - \mathrm{D}F(x_0)\|_{\mathcal{L}(X,Y^\star)}}{\|x - x_0\|_X} < \infty,$$

where $B_{R_0}(x_0) := \{x \in X \mid \|x - x_0\|_X < R_0\}$

We set

$$R := \min\{R_0, \gamma^{-1}\|\mathrm{D}F(x_0)^{-1}\|^{-1}_{\mathcal{L}(Y^\star,X)}, 2\gamma^{-1}\|\mathrm{D}F(x_0)\|_{\mathcal{L}(X,Y^\star)}\}.$$

Then the inequality

$$\frac{1}{2}\|\mathrm{D}F(x_0)\|^{-1}_{\mathcal{L}(X,Y^\star)}\|F(x)\|_{Y^\star} \le \|x - x_0\|_X \le 2\|\mathrm{D}F(x_0)^{-1}\|_{\mathcal{L}(Y^\star,X)}\|F(x)\|_{Y^\star} \tag{2.7}$$

holds for all $x \in X$ with $\|x - x_0\|_X < R$.

The proof of Lemma 2.4 is given in [104]. We repeat it for the sake of completeness in a slightly extended form, since some parts of the terminology used in [104] were not clarified there.

Proof. Since the operator $\mathrm{D}F$ is Lipschitz continuous at x_0 with the constant γ, there is a number $R_0 > 0$ such that $\|\mathrm{D}F(x) - \mathrm{D}F(x_0)\|_{\mathcal{L}(X,Y^\star)} \le \gamma\|x - x_0\|_X$ for all $x \in X$ with $\|x - x_0\|_X < R_0$. In particular, the relation

$$\|\mathrm{D}F(x_0 + t(x - x_0)) - \mathrm{D}F(x_0)\|_{\mathcal{L}(X,Y^\star)} \le \gamma\|x_0 + t(x - x_0) - x_0\|_X = \gamma t\,\|(x - x_0)\|_X \tag{2.8}$$

holds for all $x \in B(x_0, R_0)$ and some arbitrary but fixed number $t \in [0, 1]$.

For $x \in B(x_0, R)$, the mean value theorem for integration yields that

$$F(x_0) - F(x) = \int_0^1 \mathrm{D}F(x + t(x_0 - x))(x_0 - x)\,\mathrm{d}t. \tag{2.9}$$

Possibly, the meaning of the integral and the validity of (2.9) in arbitrary Hilbert spaces X and Y is not clear at once. This is why we will give a short interpretation and verification of this equality. First let $f : [0,1] \to Y^\star$ be a function which assigns a linear functional in $Y^\star$ to a number $t \in [0,1]$. The integral $\int_0^1 f(t)\,\mathrm{d}t =: y$ shall be that functional $y \in Y^\star$ which satisfies

$$\langle y, v\rangle = \int_0^1 \langle f(t), v\rangle\,\mathrm{d}t \qquad \forall v \in Y,$$

where $\langle\cdot,\cdot\rangle$ denotes the duality product, i.e. $\langle y, v\rangle = y(v)$. Next, we define for each $v \in Y$ the scalar function $\varphi_v : [0,1] \to \mathbb{C}$,

$$\varphi_v(t) := \langle F(x_0 + t(x - x_0)), v\rangle.$$

Obviously $\varphi_v(0) = \langle F(x_0), v\rangle$, $\varphi_v(1) = \langle F(x), v\rangle$ and

$$\begin{aligned}
\varphi_v'(t) &= \lim_{s\to 0} \frac{1}{s}(\varphi_v(t+s) - \varphi_v(t)) \\
&= \lim_{s\to 0} \frac{1}{s}\Big(\langle F(x_0 + t(x - x_0) + s(x - x_0)) - F(x_0 + t(x - x_0)), v\rangle\Big) \\
&= \langle \mathrm{D}F(x_0 + t(x - x_0))(x - x_0), v\rangle.
\end{aligned}$$

The calculation rules of differentiation and integration imply that the mean value theorem holds also for scalar functions which map from $\mathbb{R}$ to $\mathbb{C}$, in particular,

$$\varphi_v(1) - \varphi_v(0) = \int_0^1 \varphi_v'(t)\,\mathrm{d}t.$$

Inserting the definition of φ_v, we obtain the equality

$$\langle F(x_0) - F(x), v\rangle = \langle \int_0^1 \mathrm{D}F(x_0 + t(x - x_0))(x - x_0)\,\mathrm{d}t, v\rangle. \tag{2.10}$$

Thus, the linear functionals $F(x_0) - F(x)$ and $\int_0^1 \mathrm{D}F(x_0 + t(x - x_0))(x - x_0)\,\mathrm{d}t$ are identical and the equality (2.9) holds.

Now recall $F(x_0) = 0$, so that we can write

$$\begin{aligned}
x - x_0 &= \mathrm{D}F(x_0)^{-1}\Big\{F(x) + \mathrm{D}F(x_0)(x - x_0) - (F(x) - F(x_0))\Big\} \\
&= \mathrm{D}F(x_0)^{-1}\Big\{F(x) + \int_0^1 [\mathrm{D}F(x_0) - \mathrm{D}F(x_0 + t(x - x_0))](x - x_0)\,\mathrm{d}t\Big\}.
\end{aligned}$$

We conclude from (2.8) that

$$\|x - x_0\|_X \;\le\; \|\mathrm{D}F(x_0)^{-1}\|_{\mathcal{L}(Y^\star, X)}\Big\{\|F(x)\|_{Y^\star}$$

$$
\begin{aligned}
& + \int_0^1 \|\mathrm{D}F(x_0) - \mathrm{D}F(x_0 + t(x - x_0))\|_{\mathcal{L}(X,Y^\star)} \|x - x_0\|_X \,\mathrm{d}t \Big\} \\
\leq\ & \|\mathrm{D}F(x_0)^{-1}\|_{\mathcal{L}(Y^\star,X)} \Big\{ \|F(x)\|_{Y^\star} + \gamma \|x - x_0\|_X^2 \int_0^1 t \,\mathrm{d}t \Big\} \\
\leq\ & \|\mathrm{D}F(x_0)^{-1}\|_{\mathcal{L}(Y^\star,X)} \Big\{ \|F(x)\|_{Y^\star} + \frac{1}{2}\gamma \|x - x_0\|_X^2 \Big\} \\
\leq\ & \|\mathrm{D}F(x_0)^{-1}\|_{\mathcal{L}(Y^\star,X)} \|F(x)\|_{Y^\star} + \frac{1}{2}\|x - x_0\|_X.
\end{aligned}
$$

for all $x \in B(x_0, R)$, where the last estimate follows from

$$
\|x - x_0\|_X^2 \leq R \cdot \|x - x_0\|_X \leq \gamma^{-1} \|\mathrm{D}F(x_0)^{-1}\|_{\mathcal{L}(Y^\star,X)}^{-1} \cdot \|x - x_0\|_X.
$$

This proves

$$
\|x - x_0\|_X \leq 2\|\mathrm{D}F(x_0)^{-1}\|_{\mathcal{L}(Y^\star,X)} \|F(x)\|_{Y^\star}
$$

To prove the other direction of the asserted estimate, we use the relations $F(x_0) = 0$ and (2.10) to obtain

$$
\begin{aligned}
\langle F(x), v \rangle &= \langle \mathrm{D}F(x_0)[x - x_0], v \rangle \\
&\quad + \langle \int_0^1 [\mathrm{D}F(x_0 + t(x - x_0)) - \mathrm{D}F(x_0)][x - x_0] \,\mathrm{d}t, v \rangle
\end{aligned}
$$

for all $v \in Y$ and therefore with (2.8),

$$
\begin{aligned}
\|F(x)\|_{Y^\star} &= \sup_{\substack{v \in Y \\ \|v\|_Y = 1}} \Big| \langle F(x), v \rangle \Big| \\
&\leq \|\mathrm{D}F(x_0)\|_{\mathcal{L}(X,Y^\star)} \|x - x_0\|_X \\
&\quad + \int_0^1 \|\mathrm{D}F(x_0 + t(x - x_0)) - \mathrm{D}F(x_0)\|_{\mathcal{L}(X,Y^\star)} \|x - x_0\|_X \,\mathrm{d}t \\
&\leq \|\mathrm{D}F(x_0)\|_{\mathcal{L}(X,Y^\star)} \|x - x_0\|_X + \frac{1}{2}\gamma \|x - x_0\|_X^2 \\
&\leq 2\|\mathrm{D}F(x_0)\|_{\mathcal{L}(X,Y^\star)} \|x - x_0\|_X.
\end{aligned}
$$

The last inequality follows from $\|x - x_0\|_X \leq R \leq 2\gamma^{-1}\|\mathrm{D}F(x_0)\|_{\mathcal{L}(X,Y^\star)}$. This completes the proof. □

The terms $\|\mathrm{D}F(x_0)\|_{\mathcal{L}(X,Y^\star)}^{-1}$ and $\|\mathrm{D}F(x_0)^{-1}\|_{\mathcal{L}(Y^\star,X)}$ in the estimate of Lemma 2.4 depend only on X, Y, F and x_0; they can therefore be treated as constants and we can write

$$
\|x - x_0\|_X \sim \|F(x)\|_{Y^\star} \quad \text{for all } x \text{ with } \|x - x_0\|_X < R. \tag{2.11}
$$

Lemma 2.5. *Let X_h and Y_h be finite dimensional subspaces of X and Y, and let $F_h : X_h \to Y_h^\star$ be a continuous operator which is an approximation of F. We consider an approximate* *solution* $x_h \in X_h$ *to* $F_h(x) = 0$ *and a restriction operator* $\mathrm{R}_h \in \mathcal{L}(Y, Y_h)$. *Then, the following estimate holds:*

$$\begin{aligned} \|F(x_h)\|_{Y^\star} &\leq \|(\mathrm{Id}_Y - \mathrm{R}_h)^\# F(x_h)\|_{Y^\star} + \|\mathrm{R}_h\|_{\mathcal{L}(Y,Y_h)} \|F(x_h) - F_h(x_h)\|_{Y_h^\star} \\ &\quad + \|\mathrm{R}_h\|_{\mathcal{L}(Y,Y_h)} \|F_h(x_h)\|_{Y_h^\star}, \end{aligned}$$

where Id_Y *denotes the identity operator on* Y.

Proof. The assertion follows from $\|\mathrm{R}_h\|_{\mathcal{L}(Y,Y_h)} = \|\mathrm{R}_h^\#\|_{\mathcal{L}(Y_h^\star, Y^\star)}$ and from

$$\begin{aligned} |\langle F(x_h), v\rangle| &= |\langle F(x_h), v - \mathrm{R}_h v\rangle + \langle F(x_h) - F_h(x_h), \mathrm{R}_h v\rangle + \langle F_h(x_h), \mathrm{R}_h v\rangle| \\ &\leq \Big(\|(\mathrm{Id}_Y - \mathrm{R}_h)^\# F(x_h)\|_{Y^\star} + \|\mathrm{R}_h^\#(F(x_h) - F_h(x_h))\|_{Y^\star} \\ &\quad + \|\mathrm{R}_h^\# F_h(x_h)\|_{Y^\star} \Big) \|v\|_Y \\ &\leq \Big(\|(\mathrm{Id}_Y - \mathrm{R}_h)^\# F(x_h)\|_{Y^\star} + \|\mathrm{R}_h^\#\|_{\mathcal{L}(Y,Y_h)} \|F(x_h) - F_h(x_h)\|_{Y_h^\star} \\ &\quad + \|\mathrm{R}_h^\#\|_{\mathcal{L}(Y,Y_h)} \|F_h(x_h)\|_{Y_h^\star} \Big) \|v\|_Y \end{aligned}$$

for all $v \in Y$. □

2.3 Spectral theory for operator pencils

This section is devoted to an introduction to spectral theory for operator pencils. We will restrict ourselves to definitions which are essential for our purposes. For details and further information, we refer to [69, 48, 53]. Moreover, we will prove a relation between the multiplicity of an eigenvalue and the invertibility of the Fréchet derivative of the operator which is associated with the given eigenvalue problem. This relation will be essential in the derivation of a posteriori error estimates.

Let $\mathbb{K} = \mathbb{R}$ or $\mathbb{K} = \mathbb{C}$ and let V be a Hilbert space over $\mathbb{K}$ with the inner product $(\cdot,\cdot)_V : V \times V \to \mathbb{K}$. The specific structure of the inner product is irrelevant at this point.

We consider a pencil $\mathcal{B} : \mathbb{K} \to \mathcal{L}(V, V)$ of linear operators, that is, $\mathcal{B}(\lambda) \in \mathcal{L}(V, V)$ for all $\lambda \in \mathbb{K}$. The adjoint operator pencil is denoted by

$$\mathcal{B}^\star(\lambda) := [\mathcal{B}(\lambda)]^\star \quad \forall \lambda \in \mathbb{K}.$$

Remark 2.6. *The spectral theory can be extended to operator pencils* $\mathcal{B} : \mathbb{K} \to \mathcal{L}(V, W)$ *with* $V \neq W$, *see, for example, [53, 48]. For the sake of simplicity and having in mind the specific application, we will content ourselves with the case* $V = W$.

We say that $\lambda_0 \in \mathbb{K}$ is an *eigenvalue* of the operator pencil $\mathcal{B}(\cdot)$, if there is an element $u_0 \in V \setminus \{0\}$, so that

$$\mathcal{B}(\lambda_0) u_0 = 0.$$

Each such u_0 is called an *eigenelement* or an *eigenfunction* of $\mathcal{B}(\cdot)$ associated with λ_0. The space of all eigenelements associated with λ_0 is called the *eigenspace* corresponding to λ_0.

In accordance with [53, Definition A.3.5], we define generalized eigenelements. Suppose that λ_0 is an eigenvalue and u_0 is an associated eigenelement. The chain $(u_0, u_1, \ldots, u_k)$,

$u_i \in V$, is called a *Jordan chain* of the length $(k+1)$ for $\mathcal{B}(\cdot)$ at λ_0, if

$$\begin{aligned} \mathcal{B}(\lambda_0)u_0 &= 0, \\ \sum_{j=0}^{m} \frac{1}{j!}\frac{\mathrm{d}^j \mathcal{B}(\lambda_0)}{\mathrm{d}\lambda^j} u_{m-j} &= 0, \qquad m = 1,\ldots,k. \end{aligned} \tag{2.12}$$

The elements $u_0,\ldots,u_k$ are called *generalized eigenelements.* We say that λ_0 is a *simple* eigenvalue if each Jordan chain for $\mathcal{B}(\cdot)$ at λ_0 has the length 1 and if the corresponding eigenspace is one-dimensional.

Remark 2.7. *If $\mathcal{B}(\lambda)$ is a Fredholm operator with index $\operatorname{Ind}\mathcal{B}(\lambda) = 0$ for all $\lambda \in \mathbb{K}$, then $\dim\ker\mathcal{B}(\lambda) = \dim\ker\mathcal{B}^\star(\lambda)$ for all $\lambda \in \mathbb{K}$. This means that the operator pencils $\mathcal{B}(\cdot)$ and $\mathcal{B}^\star(\cdot)$ have the same eigenvalues and the corresponding eigenspaces have the same (finite) dimensions. In particular, λ_0 is a simple eigenvalue of $\mathcal{B}(\cdot)$ if and only if λ_0 is a simple eigenvalue of $\mathcal{B}^\star(\cdot)$. The relation (2.4) implies that any element $w \in V$ which is orthogonal (unitary) to the eigenspace of $\mathcal{B}^\star(\lambda_0)$ is in the range of $\mathcal{B}(\lambda_0)$.*

We search for eigenpairs of the operator pencils $\mathcal{B}(\cdot)$, that is, for $\lambda \in \mathbb{K}$ and $u \in V$ which satisfy

$$\mathcal{B}(\lambda)u = 0.$$

To discard the trivial solutions $u = 0$, one usually requires that $\|u\|^2 = 1$, where $\|u\|$ is any norm in V. It is not necessary that this norm is defined by $\|u\|^2 = (u,u)_V$; any other inner product on V will serve the purpose. Let $(\cdot,\cdot)_H : V \times V \to \mathbb{K}$ be a second inner product on V; the space V equipped with $(\cdot,\cdot)_H$ needs not to be complete with respect to associated norm $\|u\|_H = \sqrt{(u,u)_H}$. All we require is that there is a constant $C_{H,V} > 0$ such that

$$\|u\|_H = \sqrt{(u,u)_H} \le C_{H,V}\sqrt{(u,u)_V} = C_{H,V}\|u\|_V.$$

We remark that $(\cdot,\cdot)_V$ and $(\cdot,\cdot)_H$ may be identical; in Verfürth's book [104], however, two different inner products were chosen at this point. In order to give a general overview of the theory which includes Verfürth's example of the eigenvalue problem for the (generalized) Laplace operator, we will make this difference here as well. Moreover, this distinction simplifies the implementation. Note that the Hilbert space theory of Section 2.1 is valid only in connection with $(\cdot,\cdot)_V$.

Independently of the definition of the second inner product, the function $g(u) = \|u\|_H^2 = (u,u)_H$ is not differentiable with respect to u in a complex Hilbert space, so that the approximation theory suggested in [104] for real-valued functions is not applicable to its full extent in the complex case. Instead, we define $g(u) := (u,\bar{u})_H$. In a real Hilbert space, this definition is equivalent to $g(u) = \|u\|_H^2$. In a complex Hilbert space, however, the requirement $g(u) = (u,\bar{u})_H = 1$ makes sense only under the additional restriction that $(u,\bar{u})_H \neq 0$ for $u \neq 0$: If u is an eigenelement of $\mathcal{B}(\cdot)$ and $g(u) \neq 0$, then $g(u) = z \in \mathbb{C}\setminus\{0\}$; hence, there is a number $c \in \mathbb{C}\setminus\{0\}$ with $c^2 = z$, and the element $\tilde{u} = u/c$ satisfies $(\tilde{u},\bar{\tilde{u}})_H = (u/c,\bar{u}/\bar{c})_H = z/c^2 = 1$ and is also an eigenelement of $\mathcal{B}(\cdot)$. The condition $g(u) = 1$ is therefore legitimate if $(u,\bar{u})_H = 0$ can be excluded.

In complex Hilbert spaces, however, there are elements $u \in V\setminus\{0\}$ which satisfy $g(u) = (u,\bar{u})_H = 0$. When $u = u_R + iu_I$ is such an element, then $\|u_R\|_H = \|u_I\|_H$ and $(u_R,u_I)_H = 0$ (cf. (2.1)). In the forthcoming theory, we will require $g(u) = 1$ and accept that we discard possible eigenelements of this form. Before we explain the meaning of this restriction (see Remark 2.9 on page 25), we need further theory concerning the spectral properties of the eigenvalue problem $\mathcal{B}(\lambda)u = 0$.

Let us consider the problem: Find $\lambda \in \mathbb{K}$, $u \in V$ such that

$$\mathcal{B}(\lambda)u = 0, \qquad (u, \bar{u})_H = 1. \tag{2.13}$$

The adjoint eigenvalue problem is given by: Find $\lambda \in \mathbb{K}$, $u^\star \in V$ such that

$$\mathcal{B}^\star(\lambda)u^\star = 0, \qquad (u^\star, \bar{u}^\star)_H = 1. \tag{2.14}$$

To derive a variational formulation for problem (2.13), we define a function $F : \mathbb{K} \times V \to \mathbb{K} \times V^\star$ via the equality

$$\langle F([\lambda, u]), [\bar{\mu}, \bar{v}]\rangle = (\mathcal{B}(\lambda)u, v)_V + \bar{\mu}\left((u, \bar{u})_H - 1\right), \qquad \forall \mu \in \mathbb{K},\ v \in V. \tag{2.15}$$

The conjugate complex element $[\bar{\mu}, \bar{v}]$ in the definition of F ensures that $F([\lambda, u])$ is linear with respect to $[\mu, v]$, cf. (2.3) and the remarks thereafter. The definition (2.15) is to be understood as follows: The object $F([\lambda, u]) = [r, f] \in \mathbb{K} \times V^\star$ with $r = (u, \bar{u})_H - 1$ and $f(\bar{v}) = (\mathcal{B}(\lambda)\, u, v)_V$ is a pair of a number r and a linear functional f and not a functional itself. Thus, the term $\langle F([\lambda, u]), [\bar{\mu}, \bar{v}]\rangle$ is not a duality product in the usual sense; we call the expression $\langle [r, f], [\bar{\mu}, \bar{v}]\rangle := r\bar{\mu} + f(\bar{v})$ the *pseudo-duality product* of $[r, f] \in \mathbb{K} \times V^\star$ and $[\mu, v] \in \mathbb{K} \times V$. Nevertheless, the linearity properties with respect to $[\mu, v]$ are satisfied,

$$\begin{aligned}
\langle \alpha[r_1, f_1] + \beta[r_2, f_2], [\bar{\mu}, \bar{v}]\rangle &= (\alpha r_1 + \beta r_1)\bar{\mu} + \langle \alpha f_1 + \beta f_2, \bar{v}\rangle \\
&= \alpha\langle [r_1, f_1], [\bar{\mu}, \bar{v}]\rangle + \beta\langle [r_2, f_2], [\bar{\mu}, \bar{v}]\rangle, \\
\langle [r, f], \alpha[\bar{\mu}_1, \bar{v}_1] + \beta[\bar{\mu}_2, \bar{v}_2]\rangle &= r(\alpha\bar{\mu}_1 + \beta\bar{\mu}_2) + \langle f, \alpha\bar{v}_1 + \beta\bar{v}_2\rangle \\
&= \alpha\langle [r, f], [\bar{\mu}_1, \bar{v}_1]\rangle + \beta\langle [r, f], [\bar{\mu}_2, \bar{v}_2]\rangle,
\end{aligned}$$

so that we keep the same notation $\langle \cdot, \cdot \rangle$ as for standard duality products.

We emphasize that, in our context, F maps to the space $\mathbb{K} \times V^\star$ and *not* to the space $(\mathbb{K} \times V)^\star$. The latter notation is used in [104], where the distinction is of minor importance, because only real eigenvalues and real functions are considered. The relations

$$\mathbb{R} \times V^\star \subset \mathbb{C} \times V^\star \quad \text{and} \quad (\mathbb{C} \times V)^\star \subset (\mathbb{R} \times V)^\star$$

make clear that $\mathbb{K} \times V^\star \neq (\mathbb{K} \times V)^\star$. At most, there is a connection between the spaces $\mathbb{K}^\star \times V^\star$ and $(\mathbb{K} \times V)^\star$. While $\mathbb{K}^\star \times V^\star$ and $\mathbb{K} \times V^\star$ consist of pairs $[r, f]$ with $r \in \mathbb{K}$ (or $r \in \mathbb{K}^\star$) and $f \in V^\star$, the structure of $(\mathbb{K} \times V)^\star$ is not clear. In the proof of Theorem 2.8, however, it will be essential to know the structure of the image space of F. We remark that the space $\mathbb{K}^\star$ consists of linear functions which map from $\mathbb{K}$ to $\mathbb{K}$, this means functions which perform a simple multiplication with a (real or complex) number. Thus, the spaces $\mathbb{K}^\star$ and $\mathbb{K}$ are isomorphic.

The norm in $\mathbb{K} \times V$ shall be given by

$$\|[\lambda, u]\|_{\mathbb{K}\times V} := |\lambda| + \|u\|_V.$$

Recall $\|u\|_V^2 = (u, u)_V$ and $(u, u)_H \leq C_{H,V}^2 \|u\|_V^2$ for all $u \in V$.

The eigenvalue problem (2.13) is equivalent to the problem: Find $\lambda \in \mathbb{K}$, $u \in V$ such that

$$\langle F([\lambda, u]), [\bar{\mu}, \bar{v}]\rangle = 0 \quad \forall \mu \in \mathbb{K},\ v \in V. \tag{2.16}$$

This follows from the positive definiteness of the inner product $(\cdot, \cdot)_V$ (in particular, $(\mathcal{B}(\lambda)u, v)_V = 0$ for all $v \in V$ and thus $\bar{\mu}((u, \bar{u})_H - 1) = 0$ for all $\mu \in \mathbb{K}$).

Theorem 2.8. *Let $\mathcal{B}(\cdot) : \mathbb{K} \to \mathcal{L}(V,V)$ be a pencil of Fredholm operators with the index zero and let $[\lambda_0, u_0] \in \mathbb{K} \times V$ be a solution to problem (2.16). Assume that*

$$(\mathcal{B}'(\lambda_0)\, u_0, u_0)_V \neq 0. \tag{2.17}$$

Then, the operator $\mathrm{D}F([\lambda_0, u_0]) \in \mathcal{L}(\mathbb{K}\times V, \mathbb{K}\times V^\star)$ is a linear homeomorphism if and only if $\lambda_0 \in \mathbb{K}$ is a simple eigenvalue of problem (2.13).

Proof. Throughout the proof, let λ_0 denote an eigenvalue of problem (2.16), or equivalently of problem (2.13), and let u_0 denote a corresponding eigenelement with $(u_0, \bar{u}_0)_H = 1$. We have to show that λ_0 is a simple eigenvalue if and only if $\mathrm{D}F([\lambda_0, u_0])$ is invertible and the inverse operator is continuous. Thus, we compute the Fréchet derivative of F at $[\lambda_0, u_0]$,

$$\begin{aligned}
\langle \mathrm{D}F([\lambda_0,u_0])([\lambda,u]), [\bar{\mu},\bar{v}]\rangle &= \lim_{t\to 0}\frac{1}{t}\Big(\langle F([\lambda_0,u_0]+t[\lambda,u]) - F([\lambda_0,u_0]), [\bar{\mu},\bar{v}]\rangle\Big),\\
&= \lim_{t\to 0}\frac{1}{t}\Big((\mathcal{B}(\lambda_0+t\lambda)\,(u_0+tu), v)_V + \bar{\mu}\,((u_0+tu, \overline{u_0+tu})_H - 1)\\
&\qquad - (\mathcal{B}(\lambda_0)\,u_0, v)_V - \bar{\mu}\,((u_0,\bar{u}_0)_H - 1)\Big)\\
&= \lim_{t\to 0}\frac{1}{t}\Big(((\mathcal{B}(\lambda_0+t\lambda) - \mathcal{B}(\lambda_0))\,u_0, v)_V + t\,(\mathcal{B}(\lambda_0+t\lambda)\,u, v)_V\Big)\\
&\qquad + \lim_{t\to 0}\frac{1}{t}\Big(\bar{\mu}\,[(u_0+tu, \bar{u}_0+t\bar{u})_H - (u_0,\bar{u}_0)_H]\Big)\\
&= (\lim_{s\to 0}\frac{\lambda}{s}(\mathcal{B}(\lambda_0+s) - \mathcal{B}(\lambda_0))\,u_0, v)_V + (\mathcal{B}(\lambda_0)\,u, v)_V\\
&\qquad + \bar{\mu}\,\lim_{t\to 0}\frac{1}{t}[t(u,\bar{u}_0)_H + t(u_0,\bar{u})_H + t^2(u,\bar{u})_H]\\
&= \lambda\,(\mathcal{B}'(\lambda_0)\,u_0, v)_V + (\mathcal{B}(\lambda_0)\,u, v)_V + \bar{\mu}\,[\overline{(\bar{u}_0, u)_H} + (u_0,\bar{u})_H]\\
&= \lambda\,(\mathcal{B}'(\lambda_0)\,u_0, v)_V + (\mathcal{B}(\lambda_0)\,u, v)_V + 2\bar{\mu}\,(u_0,\bar{u})_H,
\end{aligned} \tag{2.18}$$

where the last identity follows from (2.2). One readily verifies that $\mathrm{D}F([\lambda_0, u_0])$ is a linear operator with respect to $[\lambda, u]$. The Cauchy-Schwarz inequality yields the boundedness property of $\mathrm{D}F([\lambda_0, u_0])$:

$$\begin{aligned}
\|\mathrm{D}F([\lambda_0,u_0])([\lambda,u])\|_{\mathbb{K}\times V^\star} &= \sup_{\substack{[\bar{\mu},v]\,\in\,\mathbb{K}\times V\\ \|[\bar{\mu},v]\|_{\mathbb{K}\times V}=1}} |\langle \mathrm{D}F([\lambda_0,u_0])([\lambda,u]), [\bar{\mu},\bar{v}]\rangle|\\
\leq \sup_{\substack{[\bar{\mu},v]\,\in\,\mathbb{K}\times V\\ \|[\bar{\mu},v]\|_{\mathbb{K}\times V}=1}} & |\lambda|\,|(\mathcal{B}'(\lambda_0)\,u_0, v)_V| + |(\mathcal{B}(\lambda_0)\,u, v)_V| + 2|\bar{\mu}|\,|(u_0,\bar{u})_H|\\
\leq \sup_{\substack{[\bar{\mu},v]\,\in\,\mathbb{K}\times V\\ \|[\bar{\mu},v]\|_{\mathbb{K}\times V}=1}} & |\lambda|\,\|\mathcal{B}'(\lambda_0)\,u_0\|_V\|v\|_V + \|\mathcal{B}(\lambda_0)\|_{\mathcal{L}(V,V)}\,\|u\|_V\|v\|_V + 2|\bar{\mu}|\,\|u_0\|_H\|\bar{u}\|_H.
\end{aligned}$$

Recall $\|[\lambda, u]\|_{\mathbb{K}\times V} := |\lambda| + \|u\|_V$, $\|u\|_H \leq C_{H,V}\|u\|_V$ and note that $\|\bar{u}\|_H = \|u\|_H$ for all $u \in V$. Moreover, $|\bar{\mu}| \leq 1$ and $\|v\|_V \leq 1$, since $\|[\mu, v]\|_{\mathbb{K}\times V} = 1$. The terms $\|\mathcal{B}'(\lambda_0)\,u_0\|_V$, $\|\mathcal{B}(\lambda_0)\|_{\mathcal{L}(V,V)}$ and $\|u_0\|_H$ are constant with respect to λ and u. Hence,

$$\|\mathrm{D}F([\lambda_0,u_0])([\lambda,u])\|_{\mathbb{K}\times V^\star} \leq c_1\,|\lambda| + c_2\,\|u\|_V \leq \max\{c_1,c_2\}\,\|[\lambda,u]\|_{\mathbb{K}\times V}$$

with $c_1 = \|\mathcal{B}'(\lambda_0)\,u_0\|_V$ and $c_2 = \|\mathcal{B}(\lambda_0)\|_{\mathcal{L}(V,V)} + C_{H,V}\|u_0\|_H$. Consequently $\mathrm{D}F([\lambda_0, u_0]) \in \mathcal{L}(\mathbb{K}\times V, \mathbb{K}\times V^\star)$, which means that the Fréchet derivative exists.

The remaining proof will be structured as follows: We will show first that $DF([\lambda_0, u_0])$ is injective if and only if λ_0 is a simple eigenvalue of (2.13). Then, it is sufficient to prove that $DF([\lambda_0, u_0])$ is surjective if λ_0 is a simple eigenvalue. This proves the bijectivity of $DF([\lambda_0, u_0])$. Vice versa, the bijectivity immediately entails the injectivity of $DF([\lambda_0, u_0])$, which implies that λ_0 is a simple eigenvalue.

A linear operator is injective if and only if it has a trivial kernel. The kernel of $DF([\lambda_0, u_0])$ is given by all elements $[\lambda, u] \in \mathbb{K} \times V$ which satisfy

$$\begin{aligned}
&\langle DF([\lambda_0, u_0])([\lambda, u]), [\bar{\mu}, \bar{v}]\rangle \\
&\quad = \lambda\,(\mathcal{B}'(\lambda_0)\,u_0, v)_V + (\mathcal{B}(\lambda_0)\,u, v)_V + 2\bar{\mu}\,(u_0, \bar{u})_H = 0 \qquad \forall \mu \in \mathbb{K},\ v \in V.
\end{aligned}$$

This relation implies (for $\mu = 0$) that

$$\lambda\,(\mathcal{B}'(\lambda_0)\,u_0, v)_V + (\mathcal{B}(\lambda_0)\,u, v)_V = 0 \quad \forall v \in V \tag{2.19}$$

and therefore $2\bar{\mu}(u_0, \bar{u})_H = 0$ for all $\mu \in \mathbb{K}$; consequently,

$$(u_0, \bar{u})_H = 0. \tag{2.20}$$

Suppose first that λ_0 is a simple eigenvalue. Let $[\lambda, u] \in \ker DF([\lambda_0, u_0])$. We will show indirectly that the relations (2.19) and (2.20) imply $\lambda = 0$ and $u = 0$.

If $\lambda \neq 0$ and $u \neq 0$, then

$$(\mathcal{B}'(\lambda_0)\,u_0, v)_V + (\mathcal{B}(\lambda_0)\,\frac{1}{\lambda}u, v)_V = 0 \quad \forall v \in V \tag{2.21}$$

due to (2.19). But $\frac{1}{\lambda}u$ is not an eigenelement corresponding to λ_0, since $(\mathcal{B}(\lambda_0)\,\frac{1}{\lambda}u, u_0)_V \neq 0$ due to (2.17) and (2.21). We conclude from (2.12) that $\frac{1}{\lambda}u$ is a generalized eigenelement corresponding to λ_0, which is a contradiction to the assumption that λ_0 is a simple eigenvalue. Hence $u = 0$ or $\lambda = 0$.

If $u = 0$, then $\lambda\,(\mathcal{B}'(\lambda_0)\,u_0, v)_V = 0$ for all $v \in V$ due to (2.19). We conclude from (2.17) that $\lambda = 0$.

If $\lambda = 0$, then $(\mathcal{B}(\lambda_0)\,u, v)_V = 0$ for all $v \in V$ due to (2.19), this means that u is an eigenelement corresponding to λ_0. The dimension of the eigenspace corresponding to λ_0 is one-dimensional, so that u is a scalar multiple of u_0, in particular, $u = tu_0$ with $t \in \mathbb{K}$. We conclude from (2.20) and $(u_0, \bar{u}_0)_H = 1$ that

$$0 = (u_0, \bar{u})_H = (u_0, \bar{t}\bar{u}_0)_H = t\,(u_0, \bar{u}_0)_H = t,$$

which yields $u = 0$.

This completes the proof that the kernel of $DF([\lambda_0, u_0])$ is trivial if λ_0 is a simple eigenvalue.

For the opposite direction, suppose that $DF([\lambda_0, u_0])$ has a trivial kernel, this means that the relations (2.19) and (2.20) are satisfied only for $\lambda = 0$ and $u = 0$. We will show first that there is no eigenelement corresponding to λ_0 which is linearly independent of u_0. Let us assume the contrary, i.e., let

$$(\mathcal{B}(\lambda_0)\,\tilde{u}, v)_V = 0 \quad \forall v \in V, \qquad \tilde{u} \neq 0, \qquad \tilde{u} \neq t \cdot u_0 \;\; \forall t \in \mathbb{K} \setminus \{0\}$$

for some $\tilde{u} \in V$. Then, the element $\tilde{u}_* = \tilde{u} - \overline{(u_0, \bar{\tilde{u}})_H} \cdot u_0 \neq 0$ is also an eigenelement of $\mathcal{B}(\cdot)$ corresponding to λ_0 and satisfies $(u_0, \bar{\tilde{u}}_*)_H = 0$ since $(u_0, \bar{u}_0)_H = 1$. This means that

$\tilde{u}_*$ fulfils (2.19) and (2.20) with $\lambda = 0$ and thus, $[0,0] \neq [0, \tilde{u}_*] \in \ker \mathrm{D}F([\lambda_0, u_0])$, which contradicts $\ker \mathrm{D}F([\lambda_0, u_0]) = \{[0,0]\}$. Consequently, the eigenspace corresponding to λ_0 is one-dimensional.

Now suppose that there is a generalized eigenelement $u_1 \in V$ corresponding to λ_0 (that is, the length of the Jordan chain at λ_0 is greater than 1). According to (2.12) and (2.17), the element u_1 is not a scalar multiple of u_0. The element $u_* = u_1 - \overline{(u_0, \bar{u}_1)_H} \cdot u_0 \neq 0$ satisfies $(u_0, \bar{u}_*)_H = 0$ and, due to (2.12) and (2.13),

$$
\begin{aligned}
&(\mathcal{B}'(\lambda_0)\, u_0, v)_V + (\mathcal{B}(\lambda_0)\, u_*, v)_V \\
&\quad = (\mathcal{B}'(\lambda_0)\, u_0, v)_V + (\mathcal{B}(\lambda_0)\, u_1, v)_V - \overline{(u_0, \bar{u}_1)_H}(\mathcal{B}(\lambda_0)\, u_0, v)_V \\
&\quad = 0 \quad \forall v \in V,
\end{aligned}
$$

that is, $[0,0] \neq [1, u_*] \in \ker \mathrm{D}F([\lambda_0, u_0])$ which is again a contradiction to the assumption.

This completes the proof that λ_0 is a simple eigenvalue if and only if $\mathrm{D}F([\lambda_0, u_0])$ is injective.

It remains to show that $\mathrm{D}F([\lambda_0, u_0])$ is surjective and that its inverse is continuous. Concerning the surjectivity of $\mathrm{D}F([\lambda_0, u_0])$, we show that the problem $\mathrm{D}F([\lambda_0, u_0])([\lambda, u]) = [r, f]$ is well-posed for each $[r, f] \in \mathbb{K} \times V^\star$, that is, for each $r \in \mathbb{K}$ and $f \in V^\star$, there is a number $\lambda \in \mathbb{K}$ and an element $u \in V$ such that

$$
\langle \mathrm{D}F([\lambda_0, u_0])([\lambda, u]), [\bar{\mu}, \bar{v}] \rangle = \langle [r, f], [\bar{\mu}, \bar{v}] \rangle = f(\bar{v}) + r\bar{\mu}
$$

for all $\mu \in \mathbb{K}$, $v \in V$. Recalling the structure of the $\mathrm{D}F([\lambda_0, u_0])$ from (2.18), we have to show that for all right-hand sides $[r, f] \in \mathbb{K} \times V^\star$ there is a $\lambda \in \mathbb{K}$ and an element $u \in V$ such that

$$
\lambda(\mathcal{B}'(\lambda_0)\, u_0, v)_V + (\mathcal{B}(\lambda_0)\, u, v)_V + 2\bar{\mu}\,(u_0, \bar{u})_H = f(\bar{v}) + r\bar{\mu} \qquad \forall \mu \in \mathbb{K},\ v \in V. \tag{2.22}
$$

To this end, we construct an auxiliary problem from whose solution one can immediately construct a solution to (2.22). For each $\rho \in \mathbb{K}$, we define an element $w_\rho \in V$ so that

$$
(w_\rho, v)_V = f(\bar{v}) - \rho\,(\mathcal{B}'(\lambda_0)\, u_0, v)_V \tag{2.23}
$$

holds for all $v \in V$. It follows from the Riesz representation theorem (see page 13) that w_ρ is always uniquely defined by (2.23). Suppose for the moment that there is a number $\rho \in \mathbb{K}$ so that the auxiliary problem

$$
\begin{aligned}
(\mathcal{B}(\lambda_0)\, u_{w_\rho}, v)_V &= (w_\rho, v)_V \quad \forall v \in V, \\
(u_0, \bar{u}_{w_\rho})_H &= 0,
\end{aligned} \tag{2.24}
$$

possesses a unique solution $u_{w_\rho} \in V$. Then, the pair $[\lambda, u] \in \mathbb{K} \times V$ with

$$
\lambda = \rho, \qquad u = u_{w_\rho} + \tfrac{r}{2}\, u_0 \tag{2.25}
$$

satisfies

$$
\begin{aligned}
&\langle \mathrm{D}F([\lambda_0, u_0])([\lambda, u]), [\bar{\mu}, \bar{v}] \rangle \\
&\quad = \rho\,(\mathcal{B}'(\lambda_0)\, u_0, v)_V + (\mathcal{B}(\lambda_0)\, u_{w_\rho}, v)_V + (\mathcal{B}(\lambda_0)\, \tfrac{r}{2} u_0, v)_V \\
&\qquad + 2\bar{\mu}\,(u_0, \bar{u}_{w_\rho})_H + 2\bar{\mu}\,(u_0, \tfrac{\bar{r}}{2}\, \bar{u}_0)_H \\
&\quad = \rho\,(\mathcal{B}'(\lambda_0)\, u_0, v)_V + (w_\rho, v)_V + 0 + 0 + 2\bar{\mu}\tfrac{r}{2} \\
&\quad = f(\bar{v}) + r\bar{\mu}
\end{aligned}
$$

due to (2.23) and (2.24). Thus, it remains to show that the is a number $\rho \in \mathbb{K}$ such that the auxiliary problem (2.24) possesses a unique solution indeed. To this end, we consider the adjoint operator $\mathcal{B}^\star(\lambda_0) = [\mathcal{B}(\lambda_0)]^\star$. By Remark 2.7, λ_0 is a simple eigenvalue of $\mathcal{B}^\star(\cdot)$; we denote by $u_0^\star \in V$ an associated eigenelement. Moreover, due to (2.4), an element $w \in V$ lies in the range of the operator $\mathcal{B}(\lambda_0)$ if and only if w and $u_0^\star$ are orthogonal (unitarian), that is, $(w, u_0^\star)_V = 0$. One easily checks that $(w_\rho, u_0^\star)_V = 0$ holds only if

$$\rho = f(\bar{u}_0^\star)\,(\mathcal{B}'(\lambda_0)\,u_0, u_0^\star)_V^{-1}.$$

To verify that $(\mathcal{B}'(\lambda_0)\,u_0, u_0^\star)_V \neq 0$ assume the contrary: since $u_0^\star \in \ker \mathcal{B}^\star(\lambda_0)$, we have $\mathcal{B}'(\lambda_0)u_0 \in R(\mathcal{B}(\lambda_0))$ due to (2.4). This means that there is an element $u \in V$ so that $(\mathcal{B}(\lambda_0)u, v)_V = (\mathcal{B}'(\lambda_0)u_0, v)_V$ for all $v \in V$. The assumption (2.17) implies that $\mathcal{B}(\lambda_0)u \neq 0$. Moreover,

$$(\mathcal{B}'(\lambda_0)u_0, v)_V + (\mathcal{B}(\lambda_0)(-u), v)_V = 0 \quad \forall v \in V,$$

which means that $-u$ is a generalized eigenelement of $\mathcal{B}(\cdot)$ corresponding to λ_0. This is a contradiction to the assumption that λ_0 is a simple eigenvalue, so we have to drop the assumption that $(\mathcal{B}'(\lambda_0)\,u_0, u_0^\star)_V = 0$.

Consequently, $w_\rho \in R(\mathcal{B}(\lambda_0))$, i.e., there is an element $u \in V$ with $(\mathcal{B}(\lambda_0)\,u, v)_V = (w_\rho, v)_V$ for all $v \in V$. Moreover, the element $u_* = u - \overline{(u_0, \bar{u})_H} \cdot u_0$ satisfies $(\mathcal{B}(\lambda_0)\,u_*, v)_V = (w_\rho, v)_V$ as well and, in addition, $(u_0, \bar{u}_*)_H = 0$. Assume that there is another element u_{**} satisfying (2.24). Then $(\mathcal{B}(\lambda_0)[u_* - u_{**}], v)_V = (\mathcal{B}(\lambda_0)u_* - \mathcal{B}(\lambda_0)u_{**}, v)_V = (w_\rho, v)_V - (w_\rho, v)_V = 0$ for all $v \in V$ and thus $[u_* - u_{**}] \in \ker \mathcal{B}(\lambda_0)$. Since λ_0 is a simple eigenvalue, we have that $[u_* - u_{**}] = t u_0$ for some $t \in \mathbb{K}$ and therefore $t = (u_0, \overline{u_* - u_{**}})_H = (u_0, \bar{u}_*)_H - (u_0, \bar{u}_{**})_H = 0$. We conclude that $u_* = u_{**}$, which means that problem (2.24) has a unique solution.

Finally, the continuity of the inverse operator $\mathrm{D}F([\lambda_0, u_0])^{-1}$ follows from Banach's theorem (theorem on inverse operators, see, for example, [2, 17]). □

Remark 2.9. *Lemma 2.4 and Lemma 2.5 remain valid even when the spaces X and $Y^\star$ are replaced by $\mathbb{K} \times V$ and $\mathbb{K} \times V^\star$, respectively. In the proofs, the duality products and the elements $v \in Y$ have to be replaced by pseudo-duality products (see page 21) and the terms $[\bar{\mu}, \bar{v}] \in \mathbb{K} \times V$, respectively.*

Theorem 2.8 states that $[\lambda_0, u_0]$ with $(u_0, \bar{u}_0)_H = 1$ is a regular solution to (2.16) if and only if λ_0 is a simple eigenvalue of the operator pencil $\mathcal{B}(\cdot)$.

The condition $(u_0, \bar{u}_0)_H \neq 0$ has an influence only on the constant in the upper error bound in the estimate (2.7) of Lemma 2.4, particularly on the term $\|\mathrm{D}F([\lambda_0, u_0])^{-1}\|_{\mathcal{L}(\mathbb{K}\times V^\star, \mathbb{K}\times V)}$. It ensures that the operator $\mathrm{D}F([\lambda_0, u_0])$ is invertible so that the constant in question is finite. Concerning this constant, it is irrelevant, however, to which value the term $(u, \bar{u})_H \in \mathbb{C} \setminus \{0\}$ is scaled. The suggested value 1 can be replaced by any other complex number $z \neq 0$, if the function F is redefined appropriately. When $(u_0, \bar{u}_0)_H = 0$, nothing can be said about the validity of the upper error bound. The lower error bound

$$\|F([\lambda_h, u_h])\|_{\mathbb{K}\times V^\star} \leq 2\|\mathrm{D}F([\lambda_0, \underline{u}_0])\|_{\mathcal{L}(\mathbb{K}\times V, \mathbb{K}\times V^\star)}\|[\lambda_0, \underline{u}_0] - [\lambda_h, \underline{u}_h]\|_{\mathbb{K}\times V}$$

holds nevertheless with a finite constant, when $[\lambda_h, u_h]$ is an approximation to $[\lambda_0, u_0]$, compare (2.7). Note that $(u, \bar{u})_H \neq 0$ for all $u \neq 0$ in a real Hilbert space.

According to Lemma 2.4, we need in addition that the operator $\mathrm{D}F$ is Lipschitz continuous at $[\lambda_0, u_0]$. The next lemma gives a sufficient condition for the Lipschitz continuity of $\mathrm{D}F$.

Lemma 2.10. *Let $[\lambda_0, u_0] \in \mathbb{K} \times V$ be a solution to (2.16). If there are constants $R_0 > 0$, $c_0 > 0$ such that*

$$\|\mathcal{B}'(\lambda_1)u_1 - \mathcal{B}'(\lambda_0)u_0\|_V + \|\mathcal{B}(\lambda_1) - \mathcal{B}(\lambda_0)\|_{\mathcal{L}(V,V)} \leq c_0\|[\lambda_1, u_1] - [\lambda_0, u_0]\|_{\mathbb{K}\times V} \tag{2.26}$$

for all $[\lambda_1, u_1] \in \mathbb{K} \times V$ with $\|[\lambda_1, u_1] - [\lambda_0, u_0]\|_{\mathbb{K}\times V} \leq R_0$, then the operator DF *is Lipschitz continuous at $[\lambda_0, u_0]$ with some constant γ.*

Proof. We have to show that there is a constant $\gamma > 0$ such that

$$\begin{aligned}
&\|\mathrm{DF}([\lambda_1, u_1]) - \mathrm{DF}([\lambda_0, u_0])\|_{\mathcal{L}(\mathbb{K}\times V, \mathbb{K}\times V^*)} \\
&= \sup_{\substack{[\lambda, u] \in \mathbb{K}\times V \\ \|[\lambda,u]\|_{\mathbb{K}\times V} = 1}} \ \sup_{\substack{[\mu, v] \in \mathbb{K}\times V \\ \|[\mu,v]\|_{\mathbb{K}\times V} = 1}} \left|\langle \mathrm{DF}([\lambda_1, u_1])([\lambda, u]) - \mathrm{DF}([\lambda_0, u_0])([\lambda, u]), [\bar{\mu}, \bar{v}]\rangle\right| \\
&\leq \gamma\|[\lambda_1, u_1] - [\lambda_0, u_0]\|_{\mathbb{K}\times V}.
\end{aligned}$$

Let $[\lambda, u], [\mu, v] \in \mathbb{K} \times V$ with $\|[\lambda, u]\|_{\mathbb{K}\times V} = 1$ and $\|[\mu, v]\|_{\mathbb{K}\times V} = 1$. Then $|\lambda| \leq 1$, $|\mu| \leq 1$, $\|u\|_H \leq C_{H,V}\|u\|_V \leq C_{H,V}$, $\|v\|_V \leq 1$ and

$$\begin{aligned}
&\left|\langle \mathrm{DF}([\lambda_1, u_1])([\lambda, u]) - \mathrm{DF}([\lambda_0, u_0])([\lambda, u]), [\bar{\mu}, \bar{v}]\rangle\right| \\
&= \Big|\Big(\lambda(\mathcal{B}'(\lambda_1)u_1, v)_V + (\mathcal{B}(\lambda_1)u, v)_V + 2\bar{\mu}(u_1, \bar{u})_H\Big) \\
&\quad - \Big(\lambda(\mathcal{B}'(\lambda_0)u_0, v)_V + (\mathcal{B}(\lambda_0)u, v)_V + 2\bar{\mu}(u_0, \bar{u})_H\Big)\Big| \\
&\leq |\lambda|\, \|\mathcal{B}'(\lambda_1)u_1 - \mathcal{B}'(\lambda_0)u_0\|_V\|v\|_V + \|\mathcal{B}(\lambda_1) - \mathcal{B}(\lambda_0)\|_{\mathcal{L}(V,V)}\|u\|_V\|v\|_V \\
&\quad + 2|\bar{\mu}|\, \|u_1 - u_0\|_H\|\bar{u}\|_H \\
&\leq c_0\|[\lambda_1, u_1] - [\lambda_0, u_0]\|_{\mathbb{K}\times V} + 2C_{H,V}\|u_1 - u_0\|_H \\
&\leq \gamma\|[\lambda_1, u_1] - [\lambda_0, u_0]\|_{\mathbb{K}\times V}
\end{aligned}$$

with $\gamma = c_0 + 2C_{H,V}^2$ if $\|[\lambda_1, u_1] - [\lambda_0, u_0]\|_{\mathbb{K}\times V} \leq R_0$. □

Remark 2.11. *If $\mathcal{B}(\cdot)$ is a linear polynomial in λ, that is, $\mathcal{B}(\lambda) = \mathcal{A} - \lambda\mathcal{J}$, then the condition (2.26) is always satisfied, since $\mathcal{B}'(\lambda) = -\mathcal{J}$ and thus*

$$\begin{aligned}
&\|\mathcal{B}'(\lambda_1)u_1 - \mathcal{B}'(\lambda_0)u_0\|_V + \|\mathcal{B}(\lambda_1) - \mathcal{B}(\lambda_0)\|_{\mathcal{L}(V,V)} \\
&= \|\mathcal{J}(u_0 - u_1)\|_V + \|(\lambda_0 - \lambda_1)\mathcal{J}\|_{\mathcal{L}(V,V)} \\
&\leq \|\mathcal{J}\|_{\mathcal{L}(V,V)}\|[\lambda_1, u_1] - [\lambda_0, u_0]\|_{\mathbb{K}\times V}.
\end{aligned}$$

Hence $R_0 > 0$ is arbitrary and $c_0 = \|\mathcal{J}\|_{\mathcal{L}(V,V)}$.

If $\mathcal{B}(\cdot)$ is a quadratic polynomial in λ, that is, $\mathcal{B}(\lambda) = \mathcal{K} - \lambda\mathcal{G} - \lambda^2\mathcal{M}$, then the estimate

$$\begin{aligned}
&\|\mathcal{B}'(\lambda_1)u_1 - \mathcal{B}'(\lambda_0)u_0\|_V + \|\mathcal{B}(\lambda_1) - \mathcal{B}(\lambda_0)\|_{\mathcal{L}(V,V)} \\
&= \|(2\lambda_0\mathcal{M} + \mathcal{G})u_0 - (2\lambda_1\mathcal{M} + \mathcal{G})u_1\|_V + \|(\lambda_0 - \lambda_1)\mathcal{G} + (\lambda_0^2 - \lambda_1^2)\mathcal{M}\|_{\mathcal{L}(V,V)} \\
&\leq \|2\mathcal{M}\|_{\mathcal{L}(V,V)}\|\lambda_0 u_0 - \lambda_1 u_1\|_V + \|\mathcal{G}\|_{\mathcal{L}(V,V)}\|u_0 - u_1\|_V \\
&\quad + |\lambda_0 - \lambda_1|\|\mathcal{G} + (\lambda_0 + \lambda_1)\mathcal{M}\|_{\mathcal{L}(V,V)} \\
&\leq 2\|\mathcal{M}\|_{\mathcal{L}(V,V)}\|\lambda_1(u_1 - u_0) + (\lambda_1 - \lambda_0)u_0\|_V + \|\mathcal{G}\|_{\mathcal{L}(V,V)}\|u_1 - u_0\|_V \\
&\quad + |\lambda_1 - \lambda_0|(\|\mathcal{G}\|_{\mathcal{L}(V,V)} + |\lambda_0 + \lambda_1|\|\mathcal{M}\|_{\mathcal{L}(V,V)}) \\
&\leq \Big(2|\lambda_1|\|\mathcal{M}\|_{\mathcal{L}(V,V)} + \|\mathcal{G}\|_{\mathcal{L}(V,V)}\Big)\|u_1 - u_0\|_V \\
&\quad + \Big(2\|u_0\|_V\|\mathcal{M}\|_{\mathcal{L}(V,V)} + \|\mathcal{G}\|_{\mathcal{L}(V,V)} + |\lambda_0 + \lambda_1|\|\mathcal{M}\|_{\mathcal{L}(V,V)}\Big)|\lambda_1 - \lambda_0|
\end{aligned}$$

holds. If $\|[\lambda_1, u_1] - [\lambda_0, u_0]\|_{\mathbb{K}\times V} \leq R_0$ *for some* $R_0 > 0$*, then* $|\lambda_1 - \lambda_0| \leq R_0$ *and thus* $|\lambda_1| \leq |\lambda_0| + R_0$*. Consequently, the condition (2.26) is fulfilled with the constant*

$$\begin{aligned} c_0 &= \max\Big\{2(|\lambda_0| + R_0)\|\mathcal{M}\|_{\mathcal{L}(V,V)} + \|\mathcal{G}\|_{\mathcal{L}(V,V)}, \\ &\qquad 2\|u_0\|_V\|\mathcal{M}\|_{\mathcal{L}(V,V)} + \|\mathcal{G}\|_{\mathcal{L}(V,V)} + (2|\lambda_0| + R_0)\|\mathcal{M}\|_{\mathcal{L}(V,V)}\Big\} \\ &\leq 2(\|[\lambda_0, u_0]\|_{\mathbb{K}\times V} + R_0)\|\mathcal{M}\|_{\mathcal{L}(V,V)} + \|\mathcal{G}\|_{\mathcal{L}(V,V)}. \end{aligned}$$

The larger the constant R_0 *is, the larger becomes the constant* γ *in the Lipschitz condition, and the admissible range for which the estimate in Lemma 2.4 holds becomes smaller. Thus,* R_0 *should be chosen moderately meaning that* $[\lambda_1, u_1]$ *should be a good approximation to* $[\lambda_0, u_0]$*.*

Finally, we estimate the norms $\|DF([\lambda_0, u_0])^{-1}\|_{\mathcal{L}(\mathbb{K}\times V^\star, \mathbb{K}\times V)}$ and $\|DF([\lambda_0, u_0])\|_{\mathcal{L}(\mathbb{K}\times V, \mathbb{K}\times V^\star)}$ which are involved in the error estimate in Lemma 2.4. They are of particular interest in the linear elasticity theory (see Section 4.3), when the dependence on the constants in the error estimates on the material parameters is to be determined.

Lemma 2.12. *Let* λ_0 *be a simple eigenvalue of (2.13) and let the assumptions of Theorem 2.8 be satisfied. Moreover, let* $u_0^\star$ *be the eigenfunction of the adjoint eigenvalue problem (2.14) corresponding to* λ_0 *and let* u_{w_ρ} *be the (unique) solution to problem (2.24). Note that the right-hand side of (2.24) (and therefore* u_{w_ρ}*) depends on a functional* $f \in V^\star$*. The estimates*

$$\|DF([\lambda_0, u_0])\|_{\mathcal{L}(\mathbb{K}\times V, \mathbb{K}\times V^\star)} \leq \|\mathcal{B}'(\lambda_0)\, u_0\|_V + \|\mathcal{B}(\lambda_0)\|_{\mathcal{L}(V,V)} + 2\,\|u_0\|_H$$

and

$$\|DF([\lambda_0, u_0])^{-1}([r, f])\|_{\mathbb{K}\times V} \leq \|u_0^\star\|_V |(\mathcal{B}'(\lambda_0)u_0, u_0^\star)_V^{-1}| + \frac{1}{2}\|u_0\|_V + \sup_{\substack{f\in V^\star \\ \|f\|_{V^\star}\leq 1}} \|u_{w_\rho}\|_V.$$

hold.

Proof. By definition,

$$\begin{aligned} \|DF([\lambda_0, u_0])\|_{\mathcal{L}(\mathbb{K}\times V, \mathbb{K}\times V^\star)} &= \sup_{\substack{[\lambda, u]\in\mathbb{K}\times V \\ \|[\lambda,u]\|_{\mathbb{K}\times V} = 1}} \sup_{\substack{[\mu, v]\in\mathbb{K}\times V \\ \|[\mu,v]\|_{\mathbb{K}\times V} = 1}} \Big|\langle DF([\lambda_0, u_0])([\lambda, u]), [\bar{\mu}, \bar{v}]\rangle\Big| \\ &= \sup_{\substack{[\lambda, u]\in\mathbb{K}\times V \\ \|[\lambda,u]\|_{\mathbb{K}\times V} = 1}} \sup_{\substack{[\mu, v]\in\mathbb{K}\times V \\ \|[\mu,v]\|_{\mathbb{K}\times V} = 1}} \Big|\lambda\,(\mathcal{B}'(\lambda_0)\,u_0, v)_V + (\mathcal{B}(\lambda_0)\,u, v)_V + 2\bar{\mu}\,(u_0, \bar{u})_H\Big| \\ &\leq \sup_{\substack{[\lambda, u]\in\mathbb{K}\times V \\ \|[\lambda,u]\|_{\mathbb{K}\times V} = 1}} \sup_{\substack{[\mu, v]\in\mathbb{K}\times V \\ \|[\mu,v]\|_{\mathbb{K}\times V} = 1}} |\lambda|\,\|\mathcal{B}'(\lambda_0)\,u_0\|_V\|v\|_V + \|\mathcal{B}(\lambda_0)\|_{\mathcal{L}(V,V)}\|u\|_V\|v\|_V \\ &\qquad + 2|\bar{\mu}|\,\|u_0\|_H\|\bar{u}\|_H. \end{aligned}$$

The first assertion follows from $|\lambda| \leq 1$, $|\mu| \leq 1$, and $\|v\|_V \leq 1$.

The norm of the inverse operator is defined by

$$\|DF([\lambda_0, u_0])^{-1}\|_{\mathcal{L}(\mathbb{K}\times V^\star, \mathbb{K}\times V)} = \sup_{\substack{[r, f]\in\mathbb{K}\times V^\star \\ \|[r,f]\|_{\mathbb{K}\times V^\star} = 1}} \|DF([\lambda_0, u_0])^{-1}([r, f])\|_{\mathbb{K}\times V}. \tag{2.27}$$

In Theorem 2.8, we showed that for each $[r, f] \in \mathbb{K} \times V^\star$, there is a unique pair $[\lambda, u] \in \mathbb{K} \times V$ with $DF([\lambda_0, u_0])([\lambda, u]) = [r, f]$ provided that λ_0 is a simple eigenvalue of (2.16). Thus,

$$\|DF([\lambda_0, u_0])^{-1}([r, f])\|_{\mathbb{K}\times V} = \|[\lambda, u]\|_{\mathbb{K}\times V} = |\lambda| + \|u\|_V. \tag{2.28}$$

Recalling the structure of λ and u from (2.25), we conclude that

$$|\lambda| = |\rho| = |f(\bar{u}_0^\star)\,(\mathcal{B}'(\lambda_0)u_0, u_0^\star)_V^{-1}| \quad \text{and} \quad \|u\|_V = \|u_{w_\rho} + \tfrac{r}{2}u_0\|_V, \tag{2.29}$$

where $u_0^\star$ is an eigenfunction of the adjoint eigenvalue problem (2.14) corresponding to λ_0 and where u_{w_ρ} is the solution to problem (2.24).

Before we estimate $|\lambda|$ and $\|u\|_V$, we remark that

$$\|[r, f]\|_{\mathbb{K}\times V^\star} = \sup_{\substack{[\mu, v] \in \mathbb{K} \times V \\ \|[\mu, v]\|_{\mathbb{K}\times V} = 1}} \left|\langle [r, f], [\bar{\mu}, \bar{v}]\rangle\right| = \sup_{\substack{[\mu, v] \in \mathbb{K} \times V \\ \|[\mu, v]\|_{\mathbb{K}\times V} = 1}} |r\bar{\mu} + f(\bar{v})|.$$

Choosing $\mu = 1$ and $v = 0$ in the supremum, we obtain that $\|[r, f]\|_{\mathbb{K}\times V^\star} \geq |r|$. Likewise, with $\mu = 0$, $\|[r, f]\|_{\mathbb{K}\times V^\star} \geq \|f\|_{V^\star}$. If $\|[r, f]\|_{\mathbb{K}\times V^\star} = 1$, then $|r| \leq 1$ and $\|f\|_{V^\star} \leq 1$.

We conclude that $|f(\bar{u}_0^\star)| \leq \|f\|_{V^\star}\|\bar{u}_0^\star\|_V \leq \|u_0^\star\|_V$ and $\|u_{w_\rho} + \frac{r}{2}u_0\|_V \leq \|u_{w_\rho}\|_V + \frac{1}{2}\|u_0\|_V$. Consequently, the second assertion of Lemma 2.12 follows from (2.27), (2.28) and (2.29). □

It remains to estimate the norm $\|u_{w_\rho}\|_V$. The operator $\mathcal{B}(\lambda_0)$ possesses the simple eigenvalue zero with the corresponding eigenfunction u_0 and is therefore not invertible. We presume that $\|u_{w_\rho}\|_V$ depends on the eigenvalue of $\mathcal{B}(\lambda_0)$ with smallest positive magnitude as well as on the right-hand side w_ρ of (2.24) which is given in equation (2.23). Note that we can estimate $\|w_\rho\|_V \leq 1 + |\rho|\,\|\mathcal{B}'(\lambda_0)u_0\|_V = 1 + \|u_0^\star\|_V\,|(\mathcal{B}'(\lambda_0)u_0, u_0^\star)_V^{-1}|\,\|\mathcal{B}'(\lambda_0)u_0\|_V$. An analytic verification of our presumption or an explicit estimate for $\|u_{w_\rho}\|_V$ is not known.

All in all, we derived in this section conditions under which the assumptions of Lemma 2.4 are satisfied and gave estimates for the constants which are involved in inequality (2.7).

2.4 Tensor calculus and differential geometry

In the investigation of corner singularities, we are particularly interested in eigenvalue problems which are defined on the unit sphere $\mathcal{S}^2$. In Chapter 4, we will discuss two model problems: the eigenvalue problems which are derived from the Laplace equation and the linear elasticity problem near a $3D$ corner, compare (1.4). Concerning the elasticity theory, we will use tensors for the description of the problem. Thus, the next section is concerned with a short introduction to tensor calculus followed by a rough overview of differential geometry, including the definition of differential operators and norms on the sphere, where we restrict ourselves to the terminology which is used in the next chapters.

This section was developed during many fruitful discussions with A. Meyer (Technische Universität Chemnitz) [74]. For more detailed definitions and further information, we refer to standard books on tensor calculus (including books on continuum mechanics) and differential geometry, see, for example, [29, 65, 68, 85, 86, 101]. Particularly, the notation of Parisch [85] is largely in compliance with ours. Parisch illustrates the connection between vector and tensor calculus and gives a good overview of the notation, vector and tensor products and calculation rules.

Tensors. Let $\mathcal{V}$ be a d-dimensional vector space (in general $\mathbb{R}^d$). A *tensor* is a mathematical object that is independent of any coordinate system. A *tensor of order n* can be represented by an n-dimensional array of scalars (components) when n bases $\{\mathbf{b}_i^1\}_{i=1}^d, \ldots, \{\mathbf{b}_i^n\}_{i=1}^d$ are specified. In some books, the tensor product (or dyadic product) of the (basis) vectors is indicated with the symbol $\otimes$. For the sake of compactness, we omit the symbol $\otimes$, as it is common among many authors, and write $\mathbf{b}_i^1\mathbf{b}_j^2$ instead of $\mathbf{b}_i^1 \otimes \mathbf{b}_j^2$. Note that the order of these vectors must not be changed, that is, the tensor product is not commutative.

When n bases $\{\mathbf{b}_i^j\}_{i=1}^d$, $j = 1, \ldots, n$, in $\mathcal{V}$ are specified, the tensor τ of order n can be written as

$$\tau = \sum_{i_1,\ldots,i_n=1}^{d} \tau^{i_1,\ldots,i_n} \mathbf{b}_{i_1}^1 \cdots \mathbf{b}_{i_n}^n .$$

The scalars $\tau^{i_1,\ldots,i_n}$ are called the *components* of the tensor τ. They depend on the specific choice of the bases $\{\mathbf{b}_i^j\}$, whereas the tensor τ is invariant under coordinate transformations. Usually, it is assumed that the components $\tau^{i_1,\ldots,i_n}$ are real. In the forthcoming theory, we will need complex tensors as well. Thus, we allow complex components, whereas the vector space $\mathcal{V}$ shall be real. Having in mind the three-dimensional linear elasticity problem as a special application, we require, in particular, $d = 3$ and $\mathcal{V} = \mathbb{R}^3$.

The term $\bar{\tau}$ shall denote the conjugate complex of the tensor τ meaning that all components are conjugated. The first order and second order tensors correspond to the vectors and matrices, respectively. When we speak about *vector functions*, we mean first order tensor-valued functions, unless a basis in $\mathbb{R}^d$ is specified.

Usually, tensors are developed n times into the same basis, that is, $\mathbf{b}_i^1 = \mathbf{b}_i^2 = \ldots = \mathbf{b}_i^n$, $i = 1, 2, 3$. This is useful to simplify the further calculations, when only co- and contravariant tensor bases are considered [50, 78], and makes sense in many cases, for example, to determine the tensor invariants or the components of a material tensor. This restriction is less convenient, however, when we want to define the gradient of a vector function $\underline{u}$ on the sphere. Vector functions, appear, for example, in the elasticity theory, where each component of the displacement function describes the displacement into the corresponding direction. A vector function $\underline{u}$ might be given, for example, in the (orthonormal) Cartesian basis or in the (curved) spherical basis $\{\mathbf{e}_\varphi, \mathbf{e}_\theta, \mathbf{e}_r\}$ which is defined by the coordinate lines of the spherical coordinates (φ, θ, r). For the definition of norms and the application of results for scalar functions, it is more convenient to develop $\underline{u}$ into the Cartesian basis. Moreover, the components of the material tensor are usually given only in Cartesian components (or implemented in this way), so that it makes sense to develop the stress and strain tensors into the Cartesian basis as well. The gradient of a vector function $\underline{u}$ is a two-dimensional tensor which is described by two bases, the contravariant tensor basis $\{\mathbf{g}^i\}_{i=1}^3$, which depends on the parametrization of the domain on which $\underline{u}$ is defined (see page 30 for a proper definition), and the basis which is chosen for $\underline{u}$, see (2.30). For this reason, we will distinguish between the single bases in the componentwise representation of the tensors.

We use the symbols '$\cdot$' and ':' to define products of two tensors. The product of a tensor of order n and a tensor of order m is a tensor of order $n+m-2$ for the *simple contraction* '$\cdot$' and a tensor of order $n+m-4$ for the *double contraction* ':'. The contractions are defined by building (one or two) inner products of the basis vectors which stand closest to each other and multiplying the components of the two tensors. For example, the simple contraction of two first order tensors $\tau = \sum_{i=1}^3 \tau^i \mathbf{b}_i^1$ and $\sigma = \sum_{j=1}^3 \sigma^j \mathbf{b}_j^2$ is a generalization of the standard inner product of vectors,

$$\tau \cdot \sigma = \Big(\sum_{i=1}^{3} \tau^i \mathbf{b}_i^1\Big) \cdot \Big(\sum_{j=1}^{3} \sigma^j \mathbf{b}_j^2\Big) := \sum_{i,j=1}^{3} \tau^i \sigma^j (\mathbf{b}_i^1 \cdot \mathbf{b}_j^2),$$

where the dot in the last term denotes the inner product of vectors in $\mathcal{V}$. The norm of a first order tensor τ is the number $|\tau| = \sqrt{\tau \cdot \tau}$. A second order tensor $\upsilon = \sum_{k,\ell=1}^{3} \upsilon^{k\ell} \mathbf{b}_k^3 \mathbf{b}_\ell^4$ defines a bilinear form on the space of first order tensors via

$$\tau \cdot \upsilon \cdot \sigma = \Big(\sum_{i=1}^{3} \tau^i \mathbf{b}_i^1\Big) \cdot \Big(\sum_{k,\ell=1}^{3} \upsilon^{k\ell} \mathbf{b}_k^3 \mathbf{b}_\ell^4\Big) \cdot \Big(\sum_{j=1}^{3} \sigma^j \mathbf{b}_j^2\Big) = \sum_{i,j,k,\ell=1}^{3} \tau^i \sigma^j \upsilon^{k\ell} (\mathbf{b}_i^1 \cdot \mathbf{b}_k^3)(\mathbf{b}_\ell^4 \cdot \mathbf{b}_j^2).$$

As for matrices, the *transposed tensor* $\upsilon^\top$ shall satisfy $\tau \cdot \upsilon \cdot \sigma = \sigma \cdot \upsilon^\top \cdot \tau$. Thus, the transposed tensor is given by $\upsilon^\top = \sum_{k,\ell=1}^{3} \upsilon^{k\ell} \mathbf{b}_\ell^4 \mathbf{b}_k^3$, i.e., by swapping the basis vectors. The double contraction of two second order tensors τ and σ is defined as follows:

$$\tau : \sigma = \Big(\sum_{i,j=1}^{3} \tau^{ij} \mathbf{b}_i^1 \mathbf{b}_j^2\Big) : \Big(\sum_{k,\ell=1}^{3} \sigma^{k\ell} \mathbf{b}_k^3 \mathbf{b}_\ell^4\Big) := \sum_{i,j,k,\ell=1}^{3} \tau^{ij} \sigma^{k\ell} (\mathbf{b}_j^2 \cdot \mathbf{b}_k^3)(\mathbf{b}_i^1 \cdot \mathbf{b}_\ell^4).$$

Likewise, products of higher-order tensors can be defined.

We remark that for unsymmetric tensors another double contraction can be defined, see, for example, [85]. For symmetric tensors, both contractions coincide. We restrict the notation to the ':'-sign.

Co- and contravariant tensor bases. In order to simplify the calculations, it is often useful to change the coordinate system. Each point $\underline{\mathbf{X}} \in \mathbb{R}^3$ can be represented by three parameters ξ_1, ξ_2, ξ_3,

$$\underline{\mathbf{X}} = \underline{\mathbf{X}}(\xi_1, \xi_2, \xi_3).$$

The *covariant and the contravariant tensor bases* $\{\mathbf{g}_i\}_{i=1}^3$ and $\{\mathbf{g}^i\}_{i=1}^3$ are given by

$$\mathbf{g}_i := \frac{\partial}{\partial \xi_i} \underline{\mathbf{X}} \quad \text{and} \quad \mathbf{g}_i \cdot \mathbf{g}^j = \delta_i^j, \; i,j = 1,2,3.$$

The gradient is defined via the operator

$$\nabla = \sum_{i=1}^{3} \mathbf{g}^i \frac{\partial}{\partial \xi_i}. \tag{2.30}$$

In the analysis of corner singularities, we are especially interested in a parametrization of a ball which is centered at the corner. Let $\xi_3 = r = |\underline{\mathbf{X}}|$ be the radial variable (distance from the corner). Each point $\underline{\mathbf{x}}$ on the unit sphere can be represented by the parameters ξ_1, ξ_2; we write $\underline{\mathbf{x}} = \underline{\mathbf{x}}(\xi_1, \xi_2) \in \mathcal{S}^2$. Note that $\underline{\mathbf{x}} = \underline{\mathbf{X}}/|\underline{\mathbf{X}}|$. This means, in particular, that $|\underline{\mathbf{x}}(\xi_1, \xi_2)| = 1$ and $\underline{\mathbf{X}}(\xi_1, \xi_2, r) = r \cdot \underline{\mathbf{x}}(\xi_1, \xi_2)$ for all (ξ_1, ξ_2) in a given parameter domain $\mathcal{G} \subset \mathbb{R}^2$. Moreover, $\underline{\mathbf{x}} \cdot \partial \underline{\mathbf{x}} / \partial \xi_i = 0$ for $i = 1, 2$, since $|\underline{\mathbf{x}}|^2 = \underline{\mathbf{x}} \cdot \underline{\mathbf{x}} = 1$. The covariant and contravariant tensor bases satisfy

$$\mathbf{g}_i = r \frac{\partial}{\partial \xi_i} \underline{\mathbf{x}}, \quad i = 1, 2, \qquad \mathbf{g}_i \cdot \mathbf{g}^j = \delta_i^j, \qquad \mathbf{g}_3 = \mathbf{g}^3 = \underline{\mathbf{x}}$$

and therefore $\mathbf{g}_i \cdot \mathbf{g}_3 = 0$ for $i = 1, 2$.

When we restrict our considerations to the unit sphere, that is to $r = 1$, we will use the index $\mathcal{S}$ to indicate this. In particular, we set

$$\begin{aligned}\mathbf{g}_i^{\mathcal{S}} &:= \mathbf{g}_i(\xi_1, \xi_2, 1) &= r^{-1}\mathbf{g}_i(\xi_1, \xi_2, r),\\ \mathbf{g}_{\mathcal{S}}^i &:= \mathbf{g}^i(\xi_1, \xi_2, 1) &= r\,\mathbf{g}^i(\xi_1, \xi_2, r) \qquad \text{for } i = 1, 2.\end{aligned}$$

For a point $\underline{\mathbf{x}} \in \mathcal{S}^2$ and a (scalar or vector) function u, the *spherical gradient* is given by

$$\nabla_{\mathcal{S}} u := \sum_{i=1}^{2} \mathbf{g}_{\mathcal{S}}^i \frac{\partial u}{\partial \xi_i}.$$

If u is a vector function, then $\nabla_{\mathcal{S}} u$ is a special second order tensor. Note that $(\nabla_{\mathcal{S}} u)^\top := \sum_{i=1}^{2} \partial u / \partial \xi_i \, \mathbf{g}_{\mathcal{S}}^i$ differs from $\nabla_{\mathcal{S}} u$.

Surface and line elements. The volume element $\mathrm{d}\Omega$ transforms to $|\mathbf{g}_1 \times \mathbf{g}_2|\,\mathrm{d}\xi_1\,\mathrm{d}\xi_2\,\mathrm{d}\xi_3 = r^2\,\mathrm{d}\mathcal{S}\,\mathrm{d}r$, where

$$\mathrm{d}\mathcal{S} := \left|\frac{\partial \underline{\mathbf{x}}}{\partial \xi_1} \times \frac{\partial \underline{\mathbf{x}}}{\partial \xi_1}\right| \mathrm{d}\xi_1\,\mathrm{d}\xi_2 = |\mathbf{g}_1^{\mathcal{S}} \times \mathbf{g}_2^{\mathcal{S}}|\,\mathrm{d}\xi_1\,\mathrm{d}\xi_2$$

is the *surface element* on the sphere.

A curve $\gamma \subset \mathcal{S}^2$ shall always be given in parametrized form, $\gamma = \{\underline{\mathbf{x}}(\xi_1, \xi_2) \in \mathcal{S}^2 \mid \xi_1 = \xi_1(t),\ \xi_2 = \xi_2(t),\ t \in [0, 1]\}$. The *line element* on the sphere reads

$$\mathrm{d}\sigma := |\dot{\underline{\mathbf{x}}}|\,\mathrm{d}t = |\dot{\xi}_1 \mathbf{g}_1^{\mathcal{S}} + \dot{\xi}_2 \mathbf{g}_2^{\mathcal{S}}|\,\mathrm{d}t.$$

The surface element $\mathrm{d}\Sigma$ of the curved manifold $\Gamma = \{(\underline{\mathbf{x}}, r) \in \mathbb{R}^3 \mid \underline{\mathbf{x}} \in \gamma,\ R_1 < r < R_2,\ 0 \le R_1 < R_2 \le \infty\}$ in Cartesian coordinates is related to $\mathrm{d}\sigma$ by $\mathrm{d}\Sigma = r\,\mathrm{d}\sigma\,\mathrm{d}r$, $r \in (R_1, R_2)$.

Norms on the unit sphere. Let $\omega \subset \mathcal{S}^2$ be a subset of the unit sphere and let $\gamma = \gamma(t)$ be a curve on $\mathcal{S}^2$. By analogy with the usual Sobolev norms, we define for scalar functions u

$$\begin{aligned}\|\lceil u \rceil\|_{0,\gamma} &:= \Big(\int_\gamma |u|^2\,\mathrm{d}\sigma\Big)^{1/2} = \Big(\int_0^1 |u(\xi_1(t), \xi_2(t))|^2 |\dot{\underline{\mathbf{x}}}|\,\mathrm{d}t\Big)^{1/2},\\ \|\lceil u \rceil\|_{0,\omega} &:= \Big(\int_\omega |u|^2\,\mathrm{d}\mathcal{S}\Big)^{1/2}, \qquad \lceil u \rceil_{1,\omega} := \Big(\int_\omega |\nabla_{\mathcal{S}} u|^2\,\mathrm{d}\mathcal{S}\Big)^{1/2}\\ \text{and} \qquad \|\lceil u \rceil\|_{1,\omega} &:= \Big(\frac{1}{4}\|\lceil u \rceil\|_{0,\omega}^2 + \lceil u \rceil_{1,\omega}^2\Big)^{1/2}.\end{aligned}$$

The factor $1/4$ in the definition of the norm $\|\lceil u \rceil\|_{1,\omega}$ was first introduced in the paper by Apel, Sändig and Solov'ev [7] (although accidentally placed in front of the 1-seminorm). The considerations in Chapter 4 will show that this innovative definition is quite useful and simplifies the verification of the assumptions made in Sections 2.2 and 2.3.

We use the norm symbols $\lceil \cdot \rceil$ and $\|\lceil \cdot \rceil\|$ to indicate that the spherical line and surface elements are used. The standard symbols $|\cdot|$ and $\|\cdot\|$ keep their usual meaning

$$|u|_{0,\gamma} := \int_0^1 |u(\xi_1, \xi_2)|^2 \sqrt{\dot{\xi}_1^2 + \dot{\xi}_2^2}\,\mathrm{d}t,$$

$$\|u\|_{k,\omega} := \Big(\sum_{i=0}^{k} |u|^2_{k,\omega}\Big)^{1/2}, \quad k=0,1, \qquad \text{where}$$

$$|u|^2_{0,\omega} := \int_\omega |u(\xi_1,\xi_2)|^2 \,\mathrm{d}\xi_1\,\mathrm{d}\xi_2, \qquad |u|^2_{1,\omega} := \int_\omega \Big(\Big|\frac{\partial u}{\partial \xi_1}\Big|^2 + \Big|\frac{\partial u}{\partial \xi_2}\Big|^2\Big)\,\mathrm{d}\xi_1\,\mathrm{d}\xi_2.$$

For vector functions $\underline{u}$, we do not introduce new norm symbols. We consider vector functions as first order tensors. Once a basis $\{\mathbf{b}_i\}_{i=1}^3$ in $\mathbb{R}^3$ is specified, $\underline{u} = \sum_{i=1}^3 u_i \mathbf{b}_i$ can be represented by its components u_i. It is common to choose $\{\mathbf{b}_i\}$ so that it equals the basis of the local coordinate system, that is, $\mathbf{b}_i = \mathbf{g}^i$. Alternatively, on can develop the vector function $\underline{u}$ into another basis, usually the Cartesian basis. The relation between the Cartesian and the spherical components of a vector function is given, for example, in [55, Section 3.2.1]. Concerning the implementation and the further analysis of the problem, it seems to be more convenient to use the Cartesian basis $\{\mathbf{e}_i\}_{i=1}^3$ as it was done, for instance, in [64] and [7]. We will follow this way as well and assume that u_1, u_2, u_3 are the Cartesian components of the vector function $\underline{u} = \sum_{i=1}^3 u_i \mathbf{e}_i$. Then, we can define

$$\lceil\!\lceil \underline{u} \rceil\!\rceil_{0,\gamma} := \Big(\sum_{i=1}^{3} \lceil\!\lceil u_i \rceil\!\rceil^2_{0,\gamma}\Big)^{1/2}, \qquad \lceil\!\lceil \underline{u} \rceil\!\rceil_{k,\omega} := \Big(\sum_{i=1}^{3} \lceil\!\lceil u_i \rceil\!\rceil^2_{k,\omega}\Big)^{1/2}, \quad k=0,1,$$

for curves $\gamma \subset \mathcal{S}^2$ and domains $\omega \subset \mathcal{S}^2$. Note that this componentwise representation of the norms is only possible when $\underline{u}$ is developed into a constant orthonormal basis. Otherwise, additional terms have to be inserted into the norm definitions, see [55, 75].

Finally, we introduce the (weighted) Sobolev spaces $\mathcal{H}^k(\omega)$, $k \in \{0,1\}$, as the set of all functions u for which the norm $\lceil\!\lceil u \rceil\!\rceil_{k,\omega}$ is bounded. Likewise, the space $[\mathcal{H}^k(\omega)]^3$ consists of vector functions $\underline{u}$ with bounded norm $\lceil\!\lceil \underline{u} \rceil\!\rceil_{k,\omega}$, $k=0,1$.

The (spatial) area of a domain $\omega \subset \mathcal{S}^2$ and the (spatial) length of a curve $\gamma \subset \mathcal{S}^2$ shall be denoted by $\lceil \omega \rceil := \lceil\!\lceil 1 \rceil\!\rceil^2_{0,\omega}$ and $\lceil \gamma \rceil := \lceil\!\lceil 1 \rceil\!\rceil^2_{0,\gamma}$, respectively.

The normal vector. Let $\gamma = \gamma(t) \subset \mathcal{S}^2$ be a curve on the unit sphere. For given constants $R_1, R_2 \in \mathbb{R}_+$, $R_1 < 1 < R_2$, a corresponding two-dimensional manifold in $\mathbb{R}^3$ is given by $\Gamma = \{\underline{\mathbf{X}}(t,r) \mid \underline{\mathbf{x}} = \underline{\mathbf{X}}/|\underline{\mathbf{X}}| \in \gamma,\ R_1 < r = |\underline{\mathbf{X}}| < R_2\}$. We define the *normal vector* $\mathbf{n}_\mathcal{S}(\underline{\mathbf{x}})$ at the point $\underline{\mathbf{x}}(\xi_1(t),\xi_2(t)) \in \gamma$ so that it equals the surface normal vector $\mathbf{n}$ at the point $X(t,1) = \underline{\mathbf{x}}(\xi_1(t),\xi_2(t))$ of Γ. Since $\mathbf{n}$ is given by

$$\mathbf{n} = \frac{\dfrac{\partial \underline{\mathbf{X}}}{\partial t} \times \dfrac{\partial \underline{\mathbf{X}}}{\partial r}}{\left|\dfrac{\partial \underline{\mathbf{X}}}{\partial t} \times \dfrac{\partial \underline{\mathbf{X}}}{\partial r}\right|} = \frac{r\,\dot{\underline{\mathbf{x}}} \times \underline{\mathbf{x}}}{|r\,\dot{\underline{\mathbf{x}}} \times \underline{\mathbf{x}}|} = \frac{\dot{\underline{\mathbf{x}}} \times \underline{\mathbf{x}}}{|\dot{\underline{\mathbf{x}}} \times \underline{\mathbf{x}}|}$$

at the point $\underline{\mathbf{X}}(t,r) \in \Gamma$, we have that

$$\mathbf{n}_\mathcal{S}(\underline{\mathbf{x}}) = \frac{\dot{\underline{\mathbf{x}}} \times \underline{\mathbf{x}}}{|\dot{\underline{\mathbf{x}}} \times \underline{\mathbf{x}}|} = \frac{\dot{\underline{\mathbf{x}}} \times \mathbf{g}^3}{|\dot{\underline{\mathbf{x}}} \times \mathbf{g}^3|}.$$

This means that $\mathbf{n}_\mathcal{S}(\underline{\mathbf{x}})$ is that vector in the tangential plane at the point $\underline{\mathbf{x}}$ which is orthogonal to the tangential vector $\dot{\underline{\mathbf{x}}}$. By definition, the tangential plane in the point $\underline{\mathbf{x}}$ is spanned by the vectors $\mathbf{g}_1^\mathcal{S}(\underline{\mathbf{x}})$ and $\mathbf{g}_2^\mathcal{S}(\underline{\mathbf{x}})$. Since the corresponding surface normal vector $\underline{\mathbf{x}} = \mathbf{g}_3$ equals

$\mathbf{g}^3$, the vector $\dot{\xi}_1\mathbf{g}^2_{\mathcal{S}} - \dot{\xi}_2\mathbf{g}^1_{\mathcal{S}}$ lies in the tangential plane and is orthogonal to the tangential vector $\dot{\underline{\mathbf{x}}} = \dot{\xi}_1\mathbf{g}^{\mathcal{S}}_1 + \dot{\xi}_2\mathbf{g}^{\mathcal{S}}_2$. Hence, the (spherical) normal vector equals (up to its sign)

$$\mathbf{n}_{\mathcal{S}} = \frac{\dot{\xi}_1\mathbf{g}^2_{\mathcal{S}} - \dot{\xi}_2\mathbf{g}^1_{\mathcal{S}}}{|\dot{\xi}_1\mathbf{g}^2_{\mathcal{S}} - \dot{\xi}_2\mathbf{g}^1_{\mathcal{S}}|}.$$

If the curve γ is a part of the boundary of a domain $\omega \subset \mathcal{S}^2$, the vector $\mathbf{n}_{\mathcal{S}}$ shall be the exterior normal vector.

3 Error Estimation

3.1 Finite element discretization

Let $\Omega_{\mathcal{S}} \subset \mathcal{S}^2$ be an open, connected subset of the unit sphere with the boundary $\partial\Omega_{\mathcal{S}}$. We denote the Dirichlet and the Neumann boundary parts of $\partial\Omega_{\mathcal{S}}$ by Γ_D and Γ_N, where Γ_D is closed with respect to $\partial\Omega_{\mathcal{S}}$ and $\Gamma_D \cap \Gamma_N = \emptyset$. We consider eigenvalue problems of the type (2.13), see page 21, where the solutions $\underline{u}$ are functions which are defined on $\Omega_{\mathcal{S}}$ and satisfy the Dirichlet and Neumann boundary conditions. The considerations are restricted to the weak formulation of eigenvalue problems of partial differential equations of second order, that is, $\underline{u} \in [\mathcal{H}^1(\Omega_{\mathcal{S}})]^d$, where d is the number of degrees of freedom in each point resulting from the original boundary value problem ($d = 1$ for the Laplace problem and $d = 3$ for the linear elasticity problem). For $d \neq 1$, most functions under consideration are vector functions. We will use underlined letters for all potential vector functions (also in the case $d = 1$), whereas scalar functions (like bubble functions or nodal basis functions) are not underlined.

Remark 3.1. *If $d > 1$, it is possible to define the boundary conditions for each component of the vector function $\underline{u}$ separately. In this case, we have to associate boundary parts Γ_{D_i} and Γ_{N_i} to each component u_i of $\underline{u}$. With $\Gamma_D \cap \Gamma_N = \emptyset$, we mean that $\Gamma_{D_i} \cap \Gamma_{N_i} = \emptyset$ for $i = 1, \ldots, d$. Likewise, $\underline{u} = 0$ on Γ_D means that $u_i = 0$ on Γ_{D_i} for $i = 1, \ldots, d$. To keep the notation at the minimum, we will restrict the terminology to Γ_D and Γ_N but emphasize that the presented theory is valid for the componentwise definition of the boundary conditions.*

The eigenvalue problem which is associated with the Laplace problem in the sense of (1.4) is symmetric and therefore possesses only real eigenvalues so that the eigenfunctions can be chosen from the real Hilbert space $\mathcal{H}^1(\Omega_{\mathcal{S}})$, see Section 4.2. The eigenvalue problem which is associated with the linear elasticity problem, however, is not symmetric; there are complex eigenvalues and the corresponding eigenfunctions live in the complex (or complexified) Hilbert space $[\mathcal{H}^1(\Omega_{\mathcal{S}})]^3$, see Section 4.3. Depending on the problem under consideration, we set $\mathbb{K} = \mathbb{R}$ or $\mathbb{K} = \mathbb{C}$ and define the (real or complex) space

$$V := \{\underline{u} \in [\mathcal{H}^1(\Omega_{\mathcal{S}})]^d \mid \underline{u} = 0 \text{ on } \Gamma_D\}$$

which shall be equipped with the $(\cdot,\cdot)_V$ and the corresponding norm $\|\underline{u}\|_V = \sqrt{(\underline{u},\underline{u})_V}$. We do not specify the inner product at this point, but simply assume that the space V is complete with respect to $\|\underline{u}\|_V$ and that

$$\lceil \underline{u} \rceil_{1,\Omega_{\mathcal{S}}} \lesssim \|\underline{u}\|_V \lesssim \|\lceil \underline{u} \rceil\|_{1,\Omega_{\mathcal{S}}} \qquad \forall \underline{u} \in V. \tag{3.1}$$

We will see in Sections 4.2 and 4.3 that it might be preferable to provide V with another inner product than the standard $\mathcal{H}^1$-inner product.

Furthermore, we define the norm

$$\|[\lambda, \underline{u}]\|_{\mathbb{K}\times V} := |\lambda| + \|\underline{u}\|_V.$$

Let $(\cdot,\cdot)_H : V \times V \to \mathbb{K}$ be a second inner product on V with

$$(\underline{u},\underline{u})_H \lesssim (\underline{u},\underline{u})_V$$

for all $\underline{u} \in V$. Let $\mathcal{B} : \mathbb{K} \to \mathcal{L}(V,V)$ be the operator pencil which is associated with the given eigenvalue problem and let the function $F : \mathbb{K} \times V \to \mathbb{K} \times V^\star$ be defined as in Section 2.3,

$$\langle F([\lambda,\underline{u}]),[\bar{\mu},\underline{\bar{v}}]\rangle = (\mathcal{B}(\lambda)\underline{u},\underline{v})_V + \bar{\mu}\left((\underline{u},\underline{\bar{u}})_H - 1\right), \qquad \mu \in \mathbb{K},\ \underline{v} \in V,$$

cf. (2.15). We assume that the assumptions of Section 2.3 are satisfied, in particular, that $\mathcal{B}(\lambda)$ is a Fredholm operator with the index 0 for all $\lambda \in \mathbb{K}$ and that $\mathcal{B}(\lambda)$ is differentiable with respect to $\lambda \in \mathbb{K}$.

Let us consider the eigenvalue problem: Find $\lambda \in \mathbb{K}$, $\underline{u} \in V$ such that

$$\langle F([\lambda,\underline{u}]),[\bar{\mu},\underline{\bar{v}}]\rangle = 0 \qquad \forall \mu \in \mathbb{K},\ \underline{v} \in V. \tag{3.2}$$

Our aim is to discretize problem (3.2) and to derive a posteriori error estimates for the solutions to the discretized problem, where we want to apply the results of Section 2.2. To this end, we have to formulate a finite element discretization of problem (3.2).

Let $\mathcal{T}_h$ be a triangulation of Ω_S, that is a partition of Ω_S into open (spherical) triangles T so that $\overline{\Omega}_S = \bigcup_{T\in\mathcal{T}_h} \overline{T}$ and the closures of any two elements are either disjoint or have one common vertex or one common edge. The discretization parameter h denotes the global mesh size and is related to the number N of degrees of freedom by $h^2 \sim N$.

Corresponding to the triangulation $\mathcal{T}_h$, the set of all vertices (nodes) shall be denoted by $\mathcal{N}_h$ and the set of all edges by $\mathcal{E}_h$. For each element $T \in \mathcal{T}_h$ and each edge $E \in \mathcal{E}_h$, we introduce the sets

$$\mathcal{E}(T) := \bigcup_{\substack{E'\in\mathcal{E}_h,\\ E'\subset\partial T}} \{E'\}, \qquad \mathcal{N}(T) := \bigcup_{\substack{\mathrm{x}\in\mathcal{N}_h,\\ \mathrm{x}\in\partial T}} \{\mathrm{x}\}, \qquad \mathcal{N}(E) := \bigcup_{\substack{\mathrm{x}\in\mathcal{N}_h,\\ \mathrm{x}\in\overline{E}}} \{\mathrm{x}\}.$$

The Dirichlet and Neumann boundary edges are collected in the sets $\mathcal{E}_{h,D}$ and $\mathcal{E}_{h,N}$, respectively, whereas all non-boundary edges are contained in $\mathcal{E}_{h,\Omega_S} := \mathcal{E}_h \setminus (\mathcal{E}_{h,D} \cup \mathcal{E}_{h,N})$. Likewise, $\mathcal{N}_{h,D}$ and $\mathcal{N}_{h,N}$ denote the sets of all nodes in $\mathcal{N}_h$ at the Dirichlet or Neumann boundary, respectively, and $\mathcal{N}_{h,\Omega_S} := \mathcal{N}_h \setminus (\mathcal{N}_{h,D} \cup \mathcal{N}_{h,N})$ contains all non-boundary nodes. We assume that the boundary conditions change only at nodes. In compliance with Remark 3.1, the sets $\mathcal{E}_{h,D}$ and $\mathcal{E}_{h,N}$ or $\mathcal{N}_{h,D}$ and $\mathcal{N}_{h,N}$ are defined componentwise for vector functions. This means, for example, that there are d sets $\mathcal{E}_{h,D_i}$ corresponding to the Dirichlet boundary parts Γ_{D_i} each, $i = 1,\ldots,d$. To reduce the amount of notation, we use these symbols without indices.

Finally, we define patches of elements which are adjacent to a given node $\mathrm{x} \in \mathcal{N}_h$, edge $E \in \mathcal{E}_h$ or element $T \in \mathcal{T}_h$, see Figure 3.1. We set

$$\omega_{\mathrm{x}} := \bigcup_{\substack{T'\in\mathcal{T}_h\\ \mathrm{x}\in\mathcal{N}(T')}} T', \qquad \omega_E := \bigcup_{\substack{T'\in\mathcal{T}_h\\ E\in\mathcal{E}(T')}} T', \qquad \omega_T := \bigcup_{\substack{T'\in\mathcal{T}_h\\ \mathcal{E}(T)\cap\mathcal{E}(T')\neq\emptyset}} T',$$

$$\tilde{\omega}_E := \bigcup_{\substack{T'\in\mathcal{T}_h\\ \mathcal{N}(E)\cap\mathcal{N}(T')\neq\emptyset}} T', \qquad \tilde{\omega}_T := \bigcup_{\substack{T'\in\mathcal{T}_h\\ \mathcal{N}(T)\cap\mathcal{N}(T')\neq\emptyset}} T'.$$

Based on the triangulation $\mathcal{T}_h$, we introduce the *finite element space* $V_h \subset V$ via nodal basis functions. An example for the definition of nodal basis functions on the sphere will be

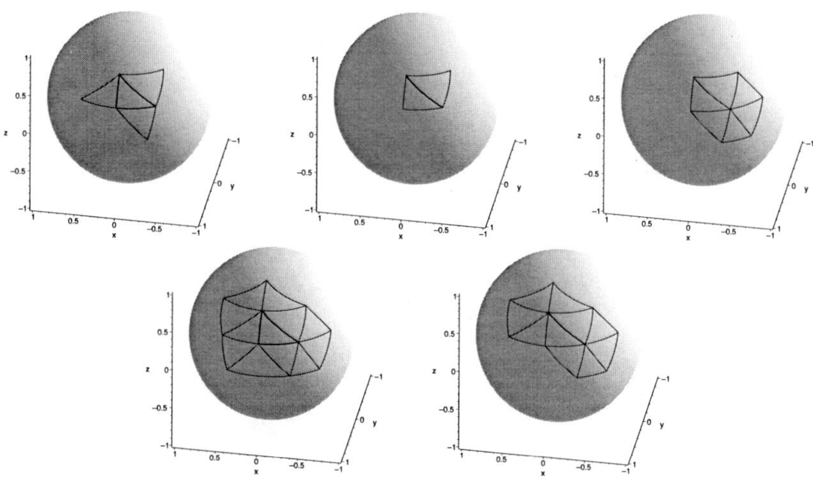

Figure 3.1: The patches ω_T, ω_E, ω_{x}, $\tilde{\omega}_T$, $\tilde{\omega}_E$ (from left top to right bottom)

given in Section 3.2 (see Remark 3.10). For the moment, let us simply assume that ϕ_{x} is the nodal basis function corresponding to the node $\mathrm{x} \in \mathcal{N}_h$ which attains the value 1 in the node x and the value 0 in all other nodes, in particular, $\operatorname{supp} \phi_{\mathrm{x}} = \overline{\omega}_{\mathrm{x}}$. Moreover, the restriction of ϕ_{x} to an element $T \in \omega_{\mathrm{x}}$ shall be a function of the two spherical parameters ξ_1, ξ_2 which is (affine) linear (or bilinear) in ξ_1 and ξ_2. We denote by Φ be the space which is spanned by all ϕ_{x} with $\mathrm{x} \in \mathcal{N}_h \setminus \mathcal{N}_{h,D}$ and set $V_h := \Phi^d \subset V$.

Remark 3.2. *If $d > 1$, then the space V_h consists of vector functions which are componentwise linear combinations of the nodal basis functions and which vanish on the Dirichlet boundary Γ_D, compare Remark 3.1.*

We define the function $F_h : \mathbb{K} \times V_h \to \mathbb{K} \times V_h^\star$ via the relation

$$\langle F_h([\lambda_h, \underline{u}_h]), [\bar{\mu}_h, \underline{\bar{v}}_h]\rangle = \langle F([\lambda_h, \underline{u}_h]), [\bar{\mu}_h, \underline{\bar{v}}_h]\rangle \qquad \forall \mu_h \in \mathbb{K},\ \forall \underline{v}_h \in V_h. \tag{3.3}$$

Note that the error $\|F([\lambda_h, \underline{u}_h]) - F_h([\lambda_h, \underline{u}_h])\|_{\mathbb{K}\times V_h^\star}$ vanishes. The *finite element discretization* of problem (3.2) is given by: Find $\lambda_h \in \mathbb{K}$, $\underline{u}_h \in V_h$ such that

$$\langle F_h([\lambda_h, \underline{u}_h]), [\bar{\mu}_h, \underline{\bar{v}}_h]\rangle = 0 \tag{3.4}$$

for all $\mu_h \in \mathbb{K}$, $\underline{v}_h \in V_h$.

3.2 Specification of the parametrization

For the verification of interpolation and approximation error estimates, further auxiliary results are needed which can be proven by considering two-dimensional analogies (for the planar case) on reference domains. To define a reference element $\hat{T}$, the parametrization of the sphere has to be chosen so that for each $T \in \mathcal{T}_h$ there is a map $\mathcal{F}_T : \hat{T} \to T$ which uniquely maps $\hat{T}$ to T. We remark that it is not necessary to restrict the considerations to only one reference element for the whole triangulation, but the number of possible reference elements has to be limited independently of the discretization.

Let ξ_1 and ξ_2 be the spherical parameters and let $\tilde{T} := \{(\xi_1, \xi_2) \in \mathbb{R}^2 \mid \underline{\mathbf{x}}(\xi_1, \xi_2) \in T\}$ be the local parameter domain corresponding to the element $T \in \mathcal{T}_h$. Note that $\tilde{T} \subset \mathbb{R}^2$ and $T \subset \mathbb{R}^3$. We assume that the domain $\tilde{T}$ has straight-lined boundary. This means, in particular, that the edges of each element can be described by $\underline{\mathbf{x}}(\xi_1(t), \xi_2(t))$, $t \in (0,1)$, so that $\dot{\xi}_1 = const$ and $\dot{\xi}_2 = const$. Then, there is an affine linear map $\tilde{\mathcal{F}}_{\tilde{T}} : \hat{T} \to \tilde{T}$ between the *reference element* $\hat{T} := \{(\hat{x}, \hat{y}) \mid 0 \leq \hat{x} \leq 1,\ 0 \leq \hat{y} \leq 1 - \hat{x}\}$ and the (planar) triangle $\tilde{T}$.

The desired parametrization can be provided, for example, by spherical coordinates (φ, θ) which are widely used, for instance, in [7, 64, 55]. The associated Jacobian is given by the term $|\mathbf{g}_1^S \times \mathbf{g}_2^S| = \sin\theta$ which can conveniently be approximated by θ in the estimates of certain integrals and norms. Moreover, spherical coordinates have the advantage that there is a global parameter domain $\tilde{\Omega}_S \subset [0, 2\pi) \times [0, \pi]$ corresponding to the entire domain Ω_S. With the transformation

$$x_1 = \cos\varphi \sin\theta, \quad x_2 = \sin\varphi \sin\theta, \quad x_3 = \cos\theta,$$

the domain $\tilde{\Omega}_S \subset \mathbb{R}^2$ is mapped to $\Omega_S = \{(x_1, x_2, x_3) \mid (\varphi, \theta) \in \tilde{\Omega}_S\} \subset \mathbb{R}^3$. If Ω_S is connected, it is possible to choose the coordinate system so that $\tilde{\Omega}_S$ is connected as well.

The drawback is that the parametrization produces artificial poles, which results in mesh crowding near the poles, when the domain Ω_S is divided equidistantly along the circles of latitude. This problem is well-known, so that a variety of alternative parametrizations is suggested in the literature, for example, to project the elements of a refined icosahedron onto the sphere [12, 40], which is common practice, for example, in applications in mathematical metrology [76, 12]. The nodes of the refined icosahedron are pulled onto the unit sphere. The element $\tilde{T}$ can be defined so that T is the projection of $\tilde{T} \subset \mathbb{R}^3$ onto the sphere, that is, $T = \{\underline{\mathbf{x}} \in \mathcal{S}^2 \mid \underline{\mathbf{x}} = \underline{\mathbf{X}}/|\underline{\mathbf{X}}|,\ \underline{\mathbf{X}} \in \tilde{T}\}$ and $\overline{T} \cap \overline{\tilde{T}} = \mathcal{N}(T)$. The corresponding reference element $\hat{T}$ can be chosen in this case as suggested before, where an affine map from $\mathbb{R}^2$ to $\mathbb{R}^3$ has to be used. This method has the drawback that it works only if $\Omega_S = \mathcal{S}^2$ or for spherical domains with geodesic boundary lines (spherical coordinates are less restrictive) and that the parameter domain is of local nature. In our applications, more general subdomains of $\mathcal{S}^2$ are considered, where the boundary of the domain Ω_S does not necessarily consist of geodesic lines. In general, there is no appropriate part of the icosahedron whose projection corresponds to Ω_S. It shall be remarked that Mu [76] used the icosahedron for the triangulation of the sphere, but spherical coordinates to define a finite element space. To avoid the difficulties at the poles, he rotated the coordinate system when necessary. This idea is worth being pursued but less practical in connection with the used meshing due to the many cases that have to be distinguished. Another alternative is a stereographic projection [38, 102], where all elements on the southern hemisphere are projected into the interior of the unit circle in the equatorial plane and all elements on the northern hemisphere to the exterior of this unit circle. This method yields a global (in general arbitrarily large) parameter domain with a quite non-uniform triangulation.

We see that each parametrization causes certain difficulties and that we have to deal with exceptions in any case. Because of the previously described advantages, we will choose spherical coordinates for the parametrization of Ω_S, see Figure 3.2. The mesh distortion which is usually caused by spherical coordinates was handled, for instance, in [63] with a so-called skipped mesh partition, where the number of grid points along a circle of latitude decreases towards the poles with a certain rate. This method produces hanging nodes and quadrilateral elements near the poles. For technical reasons, we prefer a partition of $\Omega_{\mathcal{S}}$ into (spherical) triangles without hanging nodes, but we use a similar idea for the triangulation of $\Omega_{\mathcal{S}}$.

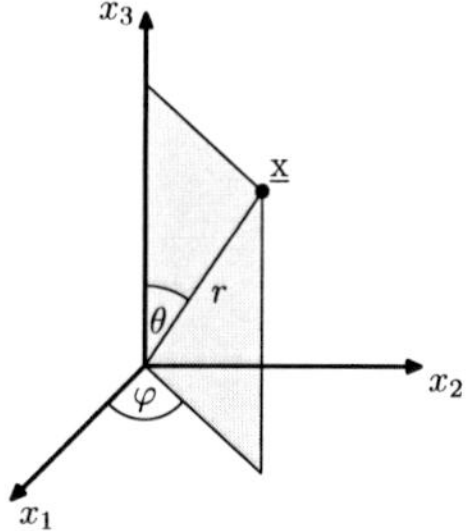

Figure 3.2: Spherical coordinates

The idea with the rotated coordinate system (see [76]) does not work when we require that the elements in the parameter domain shall have straight-lined boundary. The latter condition is quite useful, however, when we want to apply methods similar to those for two-dimensional (planar) domains.

In the following, let $\tilde{\Omega}_{\mathcal{S}}$ denote the parameter domain which is produced by spherical coordinates and corresponds to the given domain $\Omega_{\mathcal{S}}$ on which problem (3.2) is defined. We say that both $\Omega_{\mathcal{S}}$ and $\tilde{\Omega}_{\mathcal{S}}$ are *regular* domains if the boundary of $\tilde{\Omega}_{\mathcal{S}}$ is piecewise straight-lined, for simplicity, say, parallel to the coordinate axes. Note that the boundary of $\tilde{\Omega}_{\mathcal{S}}$ is not necessarily mapped to the boundary of $\Omega_{\mathcal{S}}$. Indeed, we have $\partial\Omega_{\mathcal{S}} = \emptyset$ if $\Omega_{\mathcal{S}} = \mathcal{S}^2$, whereas $\partial\tilde{\Omega}_{\mathcal{S}}$ is the boundary of the rectangle $[0, 2\pi) \times [0, \pi]$. This effect is caused by the poles, which appear as the boundary lines $\theta \equiv 0$ and $\theta \equiv \pi$ in the parameter domain, and by the fact that the lines $\varphi \equiv 0$ and $\varphi \equiv 2\pi$ are identical on the sphere. If both lines, $\varphi \equiv 0$ and $\varphi \equiv 2\pi$, appear in the parameter domain, we speak about *periodic boundary conditions* meaning that the values of the eigenfunctions have to agree on both lines. Figure 3.3 demonstrates some examples of regular domains, where the dotted lines in the parameter domains indicate the periodic boundary conditions, whereas along the thick lines, Dirichlet or Neumann boundary conditions are to be chosen.

We want to derive an a posteriori error estimator which provides both an upper and a lower bound for the error $\|[\lambda_0, \underline{u}_0] - [\lambda_h, \underline{u}_h]\|_{\mathbb{K}\times V}$, where $[\lambda_h, \underline{u}_h]$ is a solution to problem (3.4) and an approximation of a solution $[\lambda_0, \underline{u}_0]$ to the eigenvalue problem (3.2). To this end, further assumptions on the mesh are needed. In particular, we require that $\mathcal{T}_h$ is an *isotropic* triangulation of $\Omega_{\mathcal{S}}$. This means that each element $T \in \mathcal{T}_h$ is shape regular, i.e., it has equivalent (spatial) dimensions in all directions. Careful analysis of the problem reveals that one obtains a reliability–efficiency gap if the triangulation of the sphere is anisotropic.

Some notation was already introduced in Section 3.1. Before we summarize all assumptions on the mesh, the reader's attention shall be drawn to the following geometric considerations.

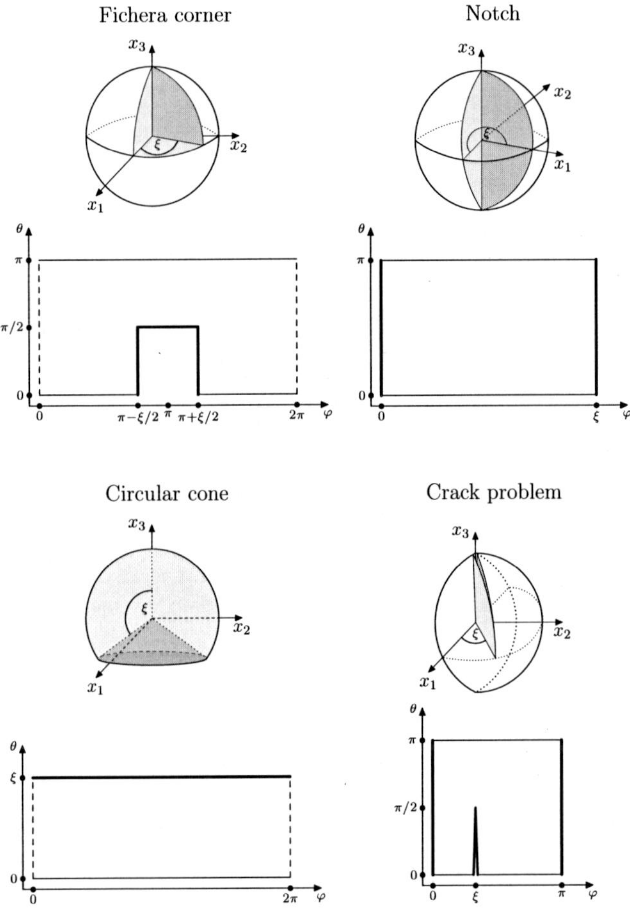

Figure 3.3: Examples of regular domains

Remark 3.3. *The spherical angles are defined by the transformation* $x_1 = \cos\varphi \sin\theta$, $x_2 = \sin\varphi \sin\theta$, $x_3 = \cos\theta$. *The term* $\sin\theta = \sqrt{x_1^2 + x_2^2}$ *equals the radius of the circle of latitude through the point* $\underline{\mathbf{x}} = (x_1, x_2, x_3) \in \mathcal{S}^2$.

The horizontal and vertical dimensions h_φ *and* h_θ *of a triangle in the parameter domain are measured by the difference of the maximum and minimum angles* φ *and* θ, *respectively. At the equator of the spherical domain, the difference* $h_\varphi = |\varphi_2 - \varphi_1|$ *of two angles* φ_1, φ_2 *equals the length of the curve enclosed by the points* $\underline{\mathbf{x}}(\varphi_1, \frac{\pi}{2})$ *and* $\underline{\mathbf{x}}(\varphi_2, \frac{\pi}{2})$. *This dimension decreases towards the pole by the factor* $\sin\theta$ *(the radius of the circle of latitude). This means that* $h_\varphi^{\mathcal{S}} := h_\varphi \sin\theta^\star$ *equals the length of the curve on the sphere which is enclosed by the two points* $\underline{\mathbf{x}}(\varphi_1, \theta^\star)$ *and* $\underline{\mathbf{x}}(\varphi_2, \theta^\star)$, *whereas* h_φ *equals the length of the line in the parameter domain with the end points* (φ_1, θ) *and* (φ_2, θ) *for arbitrary* $\theta \in [0, \pi]$.

Let $\tilde{T}$ *be an element in the parameter domain with the dimensions* $h_{\varphi,T}$ *and* $h_{\theta,T}$, *and let* $\varphi_0 = \inf_{(\varphi,\theta)\in\tilde{T}} \varphi$. *The application of the matrix*

$$\mathrm{D} = \begin{pmatrix} \sin\theta & 0 \\ 0 & 1 \end{pmatrix}$$

to $\tilde{T}$ *compresses* $\tilde{T}$ *in the* φ*-direction by the factor* $\sin\theta$, *where* $\mathrm{D}\tilde{T}$ *shall be understood as*

$$\mathrm{D}\tilde{T} := \{(\varphi_{new}, \theta_{new}) \in \mathbb{R}^2 \mid \varphi_{new} = \varphi + (\varphi - \varphi_0)\sin\theta,\ \theta_{new} = \theta,\ (\varphi, \theta) \in \tilde{T}\}. \tag{3.5}$$

The element $\mathrm{D}\tilde{T}$ *has the new dimensions* $h_{\varphi,T}^{\mathcal{S}} = h_{\varphi,T} \sup_{(\varphi,\theta)\in\tilde{T}} \sin\theta$ *and* $h_{\theta,T}^{\mathcal{S}} = h_{\theta,T}$. *These are exactly the spatial dimensions of the element* T. *Thus, it is sufficient to consider* $\mathrm{D}\tilde{T}$ *for the introduction of notation concerning the dimensions of* T. *We remark that* $\mathrm{D}\tilde{T}$ *is not a triangle in* $\mathbb{R}^2$*; the vertices are connected by curved lines.*

Definition 3.4. *For a domain* $\omega \subset \Omega_{\mathcal{S}}$ *and its counterpart* $\tilde{\omega} \subset \tilde{\Omega}_{\mathcal{S}}$, *we set*

$$\begin{aligned}
\vartheta_{+,\omega} &:= \sup_{(\varphi,\theta)\in\tilde{\omega}} \sin\theta, & \vartheta_{-,\omega} &:= \inf_{(\varphi,\theta)\in\tilde{\omega}} \sin\theta, \\
h_{\varphi,\omega} &:= \sup_{(\varphi,\theta)\in\tilde{\omega}} \varphi - \inf_{(\varphi,\theta)\in\tilde{\omega}} \varphi, & h_{\theta,\omega} &:= \sup_{(\varphi,\theta)\in\tilde{\omega}} \theta - \inf_{(\varphi,\theta)\in\tilde{\omega}} \theta, \\
h_{\varphi,\omega}^{\mathcal{S}} &:= h_{\varphi,\omega}\vartheta_{+,\omega}, & h_{\theta,\omega}^{\mathcal{S}} &:= h_{\theta,\omega}. \\
h_{+,\omega} &:= \max\{h_{\varphi,\omega}^{\mathcal{S}}, h_{\theta,\omega}^{\mathcal{S}}\}, & h_{-,\omega} &:= \min\{h_{\varphi,\omega}^{\mathcal{S}}, h_{\theta,\omega}^{\mathcal{S}}\}.
\end{aligned}$$

Furthermore, for $T \in \mathcal{T}_h$ *with* $\vartheta_{-,T} > 0$ *and the corresponding triangle* $\tilde{T}$ *in the parameter domain, let* δ_T *be the diameter of the inscribed ball of* $\mathrm{D}\tilde{T}$, *cf. (3.5).*

Assumptions on the mesh

Conformity The triangulation $\mathcal{T}_h$ of $\Omega_{\mathcal{S}}$ consists of open elements T and satisfies

$$\overline{\Omega}_{\mathcal{S}} = \bigcup_{T\in\mathcal{T}_h} \overline{T}.$$

For each pair T_1, T_2 of elements in $\mathcal{T}_h$, the intersection $\overline{T}_1 \cap \overline{T}_2$ is either empty or a common vertex or a common edge of the two elements.

Regularity The boundary of the parameter domain $\tilde{\Omega}_{\mathcal{S}}$ is piecewise parallel to the coordinate axes. The triangulation of $\Omega_{\mathcal{S}}$ is the image of a partition of $\tilde{\Omega}_{\mathcal{S}}$, where all elements in $\tilde{\Omega}_{\mathcal{S}}$ have straight-lined boundary.

A pole belongs to $\overline{\Omega}_{\mathcal{S}}$ if and only if it is contained in the set $\mathcal{N}_h$.

Adjacent elements The number of elements with a common vertex is bounded from above by a constant $\mathcal{Z}$ independently of the discretization parameter h.

The size of adjacent elements does not change rapidly. In particular, $h^{\mathcal{S}}_{\varphi,T_1} \sim h^{\mathcal{S}}_{\varphi,T_2}$ and $h^{\mathcal{S}}_{\theta,T_1} \sim h^{\mathcal{S}}_{\theta,T_2}$ for elements $T_1, T_2 \in \mathcal{T}_h$ which have at least one vertex in common.

Isotropy The triangulation $\mathcal{T}_h$ consists of *isotropic* elements, this means that each element $T \in \mathcal{T}_h$ has approximately the same (spatial) dimensions in all directions. In formulae, we require that

$$h^{\mathcal{S}}_{\varphi,T} \sim h^{\mathcal{S}}_{\theta,T} \tag{3.6}$$

and

$$\delta_T \sim h_{-,T}. \tag{3.7}$$

Pole elements Pole elements have one node at the pole and two nodes which lie on the same circle of latitude. In the parameter domain, pole elements appear as rectangles whose boundary is parallel to the φ- and θ-axes. Moreover, if $T \in \mathcal{T}_h$ is an element with $\vartheta_{-,T} > 0$ and $\vartheta_{-,\omega_T} = 0$, the associated triangle $\tilde{T}$ shall have one edge which is parallel to the φ-axis.

Sufficient fineness The mesh is fine enough so that the patches ω_T with $\vartheta_{-,\omega_T} = 0$ touch boundary corners at most directly at the pole (where $\sin\theta = 0$); crack tips must not be touched by such patches. Moreover, $h^{\mathcal{S}}_{\theta,\omega_T} \leq \frac{\pi}{4}$ for elements T with $\vartheta_{-,\omega_T} = 0$ and $h^{\mathcal{S}}_{\theta,T} \leq 1$ for all other elements.

Boundary conditions If there are nodes on the line $\varphi \equiv 2\pi$, they are identified with the nodes on the line $\varphi \equiv 0$; this means that functions on such domains satisfy *periodic boundary conditions* (they have the same values at the points $\underline{\mathbf{x}}(0,\theta)$ and $\underline{\mathbf{x}}(2\pi,\theta)$).

Elements $T \in \mathcal{T}_h$ do not cross the line $\varphi \equiv 0$ (or $\varphi \equiv 2\pi$).

Dirichlet and Neumann boundary conditions change only at nodes.

Boundary corners and crack tips are represented by nodes in $\mathcal{N}_h$.

The following algorithm was published already in [6]. It gives an example how a mesh with the desired properties can be created by producing a decomposition of the parameter domain $\tilde{\Omega}_{\mathcal{S}} \subset \mathbb{R}^2$. The description is reduced to the case $\Omega_{\mathcal{S}} = \mathcal{S}^2$. The algorithm can easily be adapted to regular subdomains of the sphere.

Algorithm 3.5. *Let* $\tilde{\Omega}_{\mathcal{S}} = [0, 2\pi) \times [0, \pi]$.

- *The parameter domain* $\tilde{\Omega}_{\mathcal{S}}$ *is divided* n_φ *times in* φ*-direction and* n_θ *times in* θ*-direction, where* n_φ *is about* 4 *or* 6 *and where* n_θ *is even.*

- *The generated rectangles at* $\theta = 0$ *and* $\theta = \pi$ *remain unchanged. They correspond to the triangles at the poles.*

  Let $h_\theta = \pi/n_\theta$. *All edges with* $\theta = k \cdot h_\theta \leq \pi/2$, $k = 1, \ldots, n_\theta/2$, *are divided equidistantly into* $k-1$ *parts.*

- *The generated nodes are connected properly with each other and the corresponding end points of the edges. The generated mesh is mirrored on the equator line* ($\theta \equiv \pi/2$), *see Figure 3.4.*

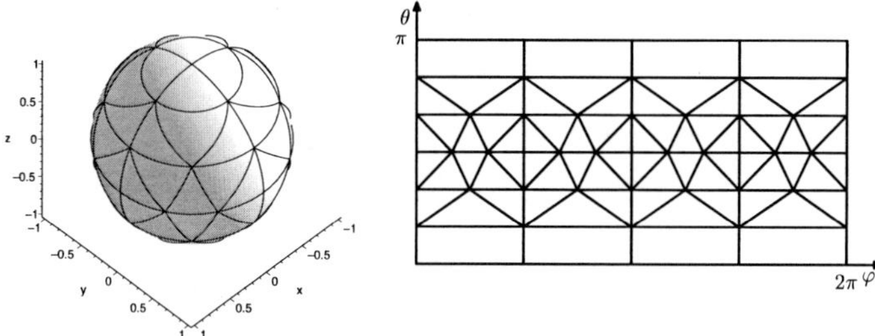

Figure 3.4: Isotropic triangulation of the sphere and the corresponding triangulation of the parameter domain according to Algorithm 3.5

The assumption that adjacent elements have approximately the same size implies that

$$h^{\mathcal{S}}_{\theta,T} \lesssim \vartheta_{-,T} \tag{3.8}$$

for all $T \in \mathcal{T}_h$ with $\vartheta_{-,T} > 0$. The following lemma is a simple consequence of this inequality.

Lemma 3.6. *([7, Section 4.3]) The relation*

$$\vartheta_{-,T} \sim \vartheta_{+,T} \tag{3.9}$$

holds for all elements $T \in \mathcal{T}_h$ with $\vartheta_{-,T} > 0$.

Proof. Relation (3.9) was proven in [7, Section 4.3]. For completeness, the proof is repeated. It is clear that $\vartheta_{+,T} \geq \vartheta_{-,T}$. It remains to show $\vartheta_{+,T} \lesssim \vartheta_{-,T}$.

By definition, there are two angles θ_- and θ_+ with $\vartheta_{-,T} = \sin\theta_-$ and $\vartheta_{+,T} = \sin\theta_+$. Obviously, $|\theta_+ - \theta_-| \leq h^{\mathcal{S}}_{\theta,T}$. The relations (3.8) and $\sin x \leq |x|$ $\forall x \in \mathbb{R}$ imply that

$$\sin(\theta_+ - \theta_-) \leq |\theta_+ - \theta_-| \leq h^{\mathcal{S}}_{\theta,T} \lesssim \vartheta_{-,T},$$

and therefore

$$\begin{aligned} \vartheta_{+,T} &= \sin\theta_+ = \sin\theta_- \cos(\theta_+ - \theta_-) + \cos\theta_- \sin(\theta_+ - \theta_-) \\ &= \vartheta_{-,T}\Big(\cos(\theta_+ - \theta_-) + \vartheta_{-,T}^{-1} \cos\theta_- \sin(\theta_+ - \theta_-)\Big) \\ &\lesssim \vartheta_{-,T}. \end{aligned}$$

□

Corollary 3.7. *Let $T \in \mathcal{T}_h$ with $\vartheta_{-,T} > 0$. Then*

$$\sin\theta \sim \vartheta_{+,T} \quad \forall \theta \textit{ with } (\varphi, \theta) \in T. \tag{3.10}$$

Remark 3.8 (pole elements). *A pole is either the point $(0,0,1)$ or $(0,0,-1)$ and can be characterized by $\sin\theta = 0$. For technical reasons, we assume that the poles belong to the set of nodes $\mathcal{N}_h$ if they are contained in $\overline{\Omega}_{\mathcal{S}}$. This ensures that for an arbitrary element $T \in \mathcal{T}_h$*

each point $\underline{\mathbf{x}} = \underline{\mathbf{x}}(\varphi, \theta) \in T$ *can be uniquely identified with the node* (φ, θ) *in the parameter domain, since we consider only open elements.*

An element $T \in \mathcal{T}_h$ *is a pole element if and only if* $\vartheta_{-,T} = 0$. *Moreover, the required shape of the pole elements implies that* $\vartheta_{+,T} = \sin h^S_{\theta,T} \sim h^S_{\theta,T}$ *and* $h^S_{\theta,T} \sim h^S_{\varphi,T} = \vartheta_{+,T} h_{\varphi,T}$. *Hence,* $h_{\varphi,T} \sim 1$, *which means that the number of pole elements is bounded and independent of the mesh parameter* h.

Remark 3.9. *The isotropy condition (3.6) entails* $h_{+,T} \sim h_{-,T}$. *Nevertheless, the condition (3.7) is not fulfilled then at once, but has to be required in addition.*

The regularity condition that the boundary of $\tilde{\Omega}_S$ *is piecewise parallel to the* φ- *and* θ-*axes can be weakened to piecewise straight-lined boundary parts. Only near the poles, the parallelity to the coordinate axes is needed in order to ensure that there is a limited number of different structures of the patches* ω_{x} *with* $\vartheta_{-,\omega_{\mathrm{x}}} = 0$.

Remark 3.10. *Let* $T \in \mathcal{T}_h$ *with the vertices* $\mathrm{x}_{i,T} = \mathrm{x}_{i,T}(\varphi_i, \theta_i) \in \mathcal{N}_h$, $i = 1, 2, 3$. *If* T *is not a pole element, the corresponding nodal basis functions can be defined over* $\tilde{T}$ *like the standard nodal basis functions for planar triangles,*

$$\begin{aligned}
\phi_{\mathrm{x}_{1,T}} &= 1 - \wp^{-1}[(\varphi - \varphi_{1,T})(\theta_{3,T} - \theta_{2,T}) - (\varphi_{3,T} - \varphi_{2,T})(\theta - \theta_{1,T})], \\
\phi_{\mathrm{x}_{2,T}} &= \wp^{-1}[(\varphi - \varphi_{1,T})(\theta_{3,T} - \theta_{1,T}) - (\varphi_{3,T} - \varphi_{1,T})(\theta - \theta_{1,T})], \\
\phi_{\mathrm{x}_{3,T}} &= \wp^{-1}[(\varphi_{2,T} - \varphi_{1,T})(\theta - \theta_{1,T}) - (\varphi - \varphi_{1,T})(\theta_{2,T} - \theta_{1,T})],
\end{aligned}$$

where $\wp = (\varphi_{2,T} - \varphi_{1,T})(\theta_{3,T} - \theta_{1,T}) - (\varphi_{3,T} - \varphi_{1,T})(\theta_{2,T} - \theta_{1,T})$.

Transformed to the reference element $\hat{T} = \{(\hat{x}, \hat{y}) \mid 0 \le \hat{x} \le 1,\ 0 \le \hat{y} \le 1 - \hat{x}\}$, *these functions correspond to*

$$\hat{\phi}_{\mathrm{x}_{1,\hat{T}}} = 1 - \hat{x} - \hat{y}, \quad \hat{\phi}_{\mathrm{x}_{2,\hat{T}}} = \hat{x}, \quad \hat{\phi}_{\mathrm{x}_{3,\hat{T}}} = \hat{y}.$$

For pole elements, let the nodes $\mathrm{x}_{1,T}$, $\mathrm{x}_{2,T}$ *and* $\mathrm{x}_{3,T}$ *be enumerated so that* $\theta_1 = \theta_2$, $\varphi_1 < \varphi_2$ *and* $\mathrm{x}_{3,T}$ *is the pole node. Hence,* $\theta_3 = 0$ *or* $\theta_3 = \pi$, *and the* φ-*coordinate* φ_3 *is not uniquely determined. We simply assume that* $\varphi_3 \in [\varphi_1, \varphi_2]$. *The corresponding nodal basis functions are (affine) bilinear in* φ *and* θ. *We set*

$$\begin{aligned}
\phi_{\mathrm{x}_{1,T}} &= \wp^{-1}(\varphi_{2,T} - \varphi)(\theta_{3,T} - \theta), \\
\phi_{\mathrm{x}_{2,T}} &= \wp^{-1}(\varphi - \varphi_{1,T})(\theta_{3,T} - \theta), \\
\phi_{\mathrm{x}_{3,T}} &= \wp^{-1}(\varphi_{2,T} - \varphi_{1,T})(\theta - \theta_{1,T})
\end{aligned}$$

with the scaling factor $\wp = (\varphi_{2,T} - \varphi_{1,T})(\theta_{3,T} - \theta_{1,T})$.

Each of these nodal basis functions is contained in $\mathcal{H}^1(\Omega_S)$ *(thus* $V_h \subset V$*) and has positive values over the corresponding element* T *with maximum value* 1 *and*

$$\sum_{\mathrm{x} \in \mathcal{N}(T)} \phi_{\mathrm{x}} = 1 \quad \textit{for all} \quad T \in \mathcal{T}_h.$$

For elements $T \in \mathcal{T}_h$ *with* $\vartheta_{-,T} > 0$, *the nodal basis functions equal the barycentric coordinates corresponding to* $\tilde{T}$. *The nodal basis functions for elements* $T \in \mathcal{T}_h$ *with* $\vartheta_{-,T} = 0$ *satisfy* $\phi_{\mathrm{x}_{1,T}} \phi_{\mathrm{x}_{2,T}} \phi_{\mathrm{x}_{3,T}} \le 1/27$, *where the maximum value* $1/27$ *is attained at* $\varphi = 1/2(\varphi_{1,T} + \varphi_{2,T})$, $\theta = 1/3(2\theta_{1,T} + \theta_{3,T})$, *and* $\phi_{\mathrm{x}_{i,T}} \phi_{\mathrm{x}_{j,T}} \le 1/4$ *for* $i \ne j$ *with maximum value* $1/4$.

3.3 Clément-type interpolation

Interpolation error estimates are essential ingredients in the proof of an upper error bound. By analogy with [6], we define a Clément-type interpolation operator. In this context, we employ (affine) linear (or bilinear) nodal basis functions. Let $T \in \mathcal{T}_h$ be an element with the nodes $\mathcal{N}(T) = \{\mathrm{x}_{1,T}, \mathrm{x}_{2,T}, \mathrm{x}_{3,T}\}$. The corresponding nodal basis functions shall be denoted by $\phi_{\mathrm{x}_{1,T}}, \phi_{\mathrm{x}_{2,T}}, \phi_{\mathrm{x}_{3,T}} : \Omega_\mathcal{S} \to \mathbb{R}$ and satisfy the standard assumptions

$$\left.\begin{array}{c} \phi_{\mathrm{x}_{i,T}}(\mathrm{x}_{j,T}) = \delta_{ij} \quad \text{and} \quad \operatorname{supp} \phi_{\mathrm{x}_{i,T}} \subseteq \overline{\omega}_{\mathrm{x}_{i,T}} \quad \text{for } i,j = 1,2,3, \\ 0 \le \phi_\mathrm{x}(\underline{\mathbf{x}}) \le 1 \quad \text{and} \quad \sum_{\mathrm{x} \in \mathcal{N}(T)} \phi_\mathrm{x}(\underline{\mathbf{x}}) = 1 \quad \forall \underline{\mathbf{x}} \in T. \end{array}\right\} \tag{3.11}$$

An example for the definition of such nodal basis functions is given in Remark 3.10.

Furthermore, let $\omega \subset \Omega_\mathcal{S}$ and let $\pi_{0,\omega} : \mathcal{H}^0(\omega) \to \mathcal{P}_0$ be the (spherical) L^2-projection of a function $v \in \mathcal{H}^0(\omega)$ into the space $\mathcal{P}_0$ of all constant functions,

$$\pi_{0,\omega} v := \frac{1}{\lceil \omega \rceil} \int_\omega v \, \mathrm{d}\mathcal{S} = \| \lceil 1 \rceil \|_{0,\omega}^{-2} \int_\omega v \, \mathrm{d}\mathcal{S}.$$

For vector functions $\underline{v} \in [\mathcal{H}^0(\omega)]^d$ which are represented in the Cartesian basis, the L^2-projection $\pi_{0,\omega} : [\mathcal{H}^0(\omega)]^d \to [\mathcal{P}_0]^d$ is defined componentwise. We do not introduce a new symbol for this L^2-projection; it will always be clear from the context, whether it operates on the space of scalar functions or on the space of vector functions.

The *Clément-type interpolation operator* is defined by

$$\mathrm{I}_h : [\mathcal{H}^0(\Omega_\mathcal{S})]^d \to V_h, \qquad \mathrm{I}_h \, \underline{v}(\underline{\mathbf{x}}) = \sum_{\mathrm{x} \in \mathcal{N}_h \setminus \mathcal{N}_{h,D}} (\pi_{0,\omega_\mathrm{x}} \underline{v}) \, \phi_\mathrm{x}(\underline{\mathbf{x}}).$$

Due to the assumptions in (3.11), constant vector functions are interpolated exactly over elements which do not touch the Dirichlet boundary, that is, $\mathrm{I}_h \, \underline{v}|_T = \underline{v}|_T$ for all $\underline{v} \in [\mathcal{P}_{0|\omega_T}]^d$ and $T \in \mathcal{T}_h$ with $\mathcal{N}(T) \cap \mathcal{N}_{h,D} = \emptyset$. Moreover, $\mathrm{I}_h \, \underline{v} = 0$ on Γ_D.

We refer again to Remark 3.1 on page 35 and emphasize that the above definition of the interpolation operator is a compact notation for

$$\mathrm{I}_h \, \underline{v}(\underline{\mathbf{x}}) = \sum_{i=1}^{d} \sum_{\mathrm{x} \in \mathcal{N}_h \setminus \mathcal{N}_{h,D_i}} (\pi_{0,\omega_\mathrm{x}} v_i) \, \phi_\mathrm{x}(\underline{\mathbf{x}}) \mathbf{e}_i.$$

Theorem 3.11 (Interpolation Error Estimates). *The interpolation operator satisfies*

$$\begin{aligned} \| \lceil \underline{v} - \mathrm{I}_h \, \underline{v} \rceil \|_{0,T} &\lesssim h_{+,T} \lceil \underline{v} \rceil_{1,\tilde{\omega}_T} \quad \forall \underline{v} \in [\mathcal{H}^1(\tilde{\omega}_T)]^d \cap V, \\ \| \lceil \underline{v} - \mathrm{I}_h \, \underline{v} \rceil \|_{0,E} &\lesssim \frac{\lceil E \rceil^{1/2}}{\lceil T \rceil^{1/2}} h_{+,T} \lceil \underline{v} \rceil_{1,\tilde{\omega}_E} \quad \forall \underline{v} \in [\mathcal{H}^1(\tilde{\omega}_E)]^d \cap V. \end{aligned}$$

The proof of Theorem 3.11 is a simple consequence of the proof for scalar functions on the sphere which is formulated in [6]. In this work, a modified proof will be presented, where some ideas from [6] and [18] are combined so that, for instance, the boundedness of the interpolation operator needs not to be shown, which simplifies the verification of further ingredients (like a Bramble-Hilbert-type lemma) considerably. The following two lemmas are essential components of the proof of Theorem 3.11.

Lemma 3.12 (Trace theorem for functions on the sphere). *Let $T \in \mathcal{T}_h$ and $E \in \mathcal{E}_h(T)$. Each $\underline{v} \in [\mathcal{H}^1(T)]^d$ satisfies*

$$\|\underline{v}\|_{0,E} \lesssim \frac{\lceil E \rceil^{1/2}}{\lceil T \rceil^{1/2}} \left(\|\underline{v}\|_{0,T}^2 + h_{+,T}^2 \lceil \underline{v} \rceil_{1,T}^2 \right)^{1/2}.$$

Lemma 3.13 (Poincaré-type inequality). *Let $\mathrm{x} \in \mathcal{N}_h$ and $T \in \omega_{\mathrm{x}}$. The relation*

$$\|\underline{v} - \pi_{0,\omega_{\mathrm{x}}} \underline{v}\|_{0,T} \leq \|\underline{v} - \pi_{0,\omega_{\mathrm{x}}} \underline{v}\|_{0,\omega_{\mathrm{x}}} \lesssim h_{+,T} \lceil \underline{v} \rceil_{1,\omega_{\mathrm{x}}}$$

holds for all $\underline{v} \in [\mathcal{H}^1(\omega_{\mathrm{x}})]^d$.

Lemma 3.13 was proven for Cartesian domains in [20, Lemma 4.3.14], where the result is called Friedrichs' inequality. Usually, this estimate is referred to as Poincaré inequality, whereas Friedrichs' inequality denotes another estimate (see Lemma 3.14).

For scalar functions on the sphere, the assertions of Lemma 3.12 and Lemma 3.13 were proven in [6]. Lemma 3.13 was stated there in a more general form. Because of the simplification of the proof of Theorem 3.11 (which will be presented in Section 3.8.2), the above form is sufficient. The idea is to transform the given domain to a reference domain in $\mathbb{R}^2$ and to apply appropriate estimates on the reference domain. To verify the Poincaré inequality, a Bramble-Hilbert-type lemma is used, where averaged Taylor polynomials of $\underline{v}$ are employed, cf. [20]. For completeness, the modified proofs of Theorem 3.11, Lemma 3.12 and Lemma 3.13 are presented in this work. Because of their complexity, they are postponed to Section 3.8.

Another auxiliary result is the following Friedrichs-type inequality. Its proof is based on the same ideas as the proofs of the previous Lemmas.

Lemma 3.14 (Friedrichs-type inequality). *Let $T \in \mathcal{T}_h$ and let $[\underline{v} \in \mathcal{H}^1(T)]^d$ vanish on a subset of $\partial\Omega$ with positive surface measure. Then,*

$$\|\underline{v}\|_{0,T} \lesssim h_{+,T} \lceil \underline{v} \rceil_{1,T}.$$

3.4 An upper error bound

The aim of this section is to derive a residual a posteriori error estimator for the solutions to the eigenvalue problem (3.2) and to prove its reliability. In the following, let $[\lambda_h, \underline{u}_h] \in \mathbb{K} \times V_h$ denote a solution to (3.4), in particular, $(\underline{u}_h, \bar{\underline{u}}_h)_H = 1$. Depending on the triangulation $\mathcal{T}_h$ of Ω_S and the discretized eigenvalue problem (3.4), there are element and edge residuals $\underline{R}_T = \underline{R}_T(\lambda_h, \underline{u}_h)$, $\underline{R}_E = \underline{R}_E(\lambda_h, \underline{u}_h)$ and $\underline{R}_N = \underline{R}_N(\lambda_h, \underline{u}_h)$ so that the function F at $[\lambda_h, \underline{u}_h]$ can be written in the form

$$\langle F([\lambda_h, \underline{u}_h]), [\bar{\mu}, \underline{v}] \rangle = \sum_{T \in \mathcal{T}_h} \int_T \underline{R}_T \cdot \underline{v} \, \mathrm{d}\mathcal{S} - \sum_{E \in \mathcal{E}_{h,\Omega_S}} \int_E \underline{R}_E \cdot \underline{v} \, \mathrm{d}\sigma + \sum_{E \in \mathcal{E}_{h,N}} \int_E \underline{R}_N \cdot \underline{v} \, \mathrm{d}\sigma, \tag{3.12}$$

see Sections 4.2 and 4.3 for examples. The Cauchy-Schwarz inequality yields for all $\mu \in \mathbb{K}$, $\underline{v} \in V$ that

$$\begin{aligned} &|\langle F([\lambda_h, \underline{u}_h]), [\bar{\mu}, \bar{\underline{v}}] \rangle| \\ &\quad \leq \sum_{T \in \mathcal{T}_h} \|\underline{R}_T\|_{0,T} \|\underline{v}\|_{0,T} + \sum_{E \in \mathcal{E}_{h,\Omega_S}} \|\underline{R}_E\|_{0,E} \|\underline{v}\|_{0,E} + \sum_{E \in \mathcal{E}_{h,N}} \|\underline{R}_N\|_{0,E} \|\underline{v}\|_{0,E} \end{aligned}$$

$$= \sum_{T\in\mathcal{T}_h} \Big\{ \lceil\!\lceil \underline{R}_T \rceil\!\rceil_{0,T} \, \lceil\!\lceil \underline{v} \rceil\!\rceil_{0,T} + \frac{1}{2} \sum_{E\in\mathcal{E}(T)\cap\mathcal{E}_{h,\Omega_S}} \lceil\!\lceil \underline{R}_E \rceil\!\rceil_{0,E} \, \lceil\!\lceil \underline{v} \rceil\!\rceil_{0,E} + \sum_{E\in\mathcal{E}(T)\cap\mathcal{E}_{h,N}} \lceil\!\lceil \underline{R}_N \rceil\!\rceil_{0,E} \, \lceil\!\lceil \underline{v} \rceil\!\rceil_{0,E} \Big\}.$$

Replacing $\underline{v}$ by $\underline{v} - \mathrm{I}_h \underline{v} \in V$ and applying Theorem 3.11, we obtain that

$$\begin{aligned}
&|\langle F([\lambda_h, \underline{u}_h]), [\bar{\mu}, \bar{\underline{v}} - \mathrm{I}_h \bar{\underline{v}}] \rangle| \\
&\lesssim \sum_{T\in\mathcal{T}_h} \Big\{ \lceil\!\lceil \underline{R}_T \rceil\!\rceil_{0,T} \cdot h_{+,T} \lceil \underline{v} \rceil_{1,\tilde{\omega}_T} + \frac{1}{2} \sum_{E\in\mathcal{E}(T)\cap\mathcal{E}_{h,\Omega_S}} \lceil\!\lceil \underline{R}_E \rceil\!\rceil_{0,E} \cdot \frac{\lceil E \rceil^{1/2}}{\lceil T \rceil^{1/2}} h_{+,T} \lceil \underline{v} \rceil_{1,\tilde{\omega}_E} \\
&\qquad + \sum_{E\in\mathcal{E}(T)\cap\mathcal{E}_{h,N}} \lceil\!\lceil \underline{R}_N \rceil\!\rceil_{0,E} \cdot \frac{\lceil E \rceil^{1/2}}{\lceil T \rceil^{1/2}} h_{+,T} \lceil \underline{v} \rceil_{1,\tilde{\omega}_E} \Big\} \\
&\le \sum_{T\in\mathcal{T}_h} \Big\{ h_{+,T}^2 \lceil\!\lceil \underline{R}_T \rceil\!\rceil_{0,T}^2 + \frac{1}{2} \sum_{E\in\mathcal{E}(T)\cap\mathcal{E}_{h,\Omega_S}} \frac{\lceil E \rceil}{\lceil T \rceil} h_{+,T}^2 \lceil\!\lceil \underline{R}_E \rceil\!\rceil_{0,E}^2 \\
&\qquad + \sum_{E\in\mathcal{E}(T)\cap\mathcal{E}_{h,N}} \frac{\lceil E \rceil}{\lceil T \rceil} h_{+,T}^2 \lceil\!\lceil \underline{R}_N \rceil\!\rceil_{0,E}^2 \Big\}^{1/2} \cdot \\
&\quad \cdot \sum_{T\in\mathcal{T}_h} \Big\{ \lceil \underline{v} \rceil_{1,\tilde{\omega}_T}^2 + \frac{1}{2} \sum_{E\in\mathcal{E}(T)\cap\mathcal{E}_{h,\Omega_S}} \lceil \underline{v} \rceil_{1,\tilde{\omega}_E}^2 + \sum_{E\in\mathcal{E}(T)\cap\mathcal{E}_{h,N}} \lceil \underline{v} \rceil_{1,\tilde{\omega}_E}^2 \Big\}^{1/2}.
\end{aligned}$$

We remark that upto the last estimate, all results are still valid with $h_{+,T} \lceil \underline{v} \rceil_{1,\omega}$ replaced by

$$\Big((h_{\varphi,T}^{\mathcal{S}})^2 \lceil\!\lceil \sin^{-1}\theta \partial \underline{v} / \partial\varphi \rceil\!\rceil_{0,\omega}^2 + (h_{\theta,T}^{\mathcal{S}})^2 \lceil\!\lceil \partial \underline{v} / \partial\theta \rceil\!\rceil_{0,\omega}^2 \Big)^{1/2},$$

in particular, in the right-hand sides of the estimates of Theorem 3.11, Lemma 3.12, Lemma 3.13 and Lemma 3.14. In order to obtain the previous estimate, the maximum of the two dimensions of an element T is needed. We will see that we need the minimum of the two dimensions to obtain a lower bound for the error as well. This is the reason, why we require the isotropy of the triangulation, that is, $h_{+,T} \sim h_{-,T}$ or, equivalently, $h_{\varphi,T}^{\mathcal{S}} \sim h_{\theta,T}^{\mathcal{S}}$, see Section 3.2.

We assumed that the number of elements in a patch $\tilde{\omega}_T$ or $\tilde{\omega}_E$ is bounded from above by a constant independent of the triangulation. Thus, the $\underline{v}$-terms in the previous inequality can be estimated by

$$\sum_{T\in\mathcal{T}_h} \Big\{ \lceil \underline{v} \rceil_{1,\tilde{\omega}_T}^2 + \frac{1}{2} \sum_{E\in\mathcal{E}(T)\cap\mathcal{E}_{h,\Omega_S}} \lceil \underline{v} \rceil_{1,\tilde{\omega}_E}^2 + \sum_{E\in\mathcal{E}(T)\cap\mathcal{E}_{h,N}} \lceil \underline{v} \rceil_{1,\tilde{\omega}_E}^2 \Big\}^{1/2} \lesssim \lceil \underline{v} \rceil_{1,\Omega_S} \lesssim \lceil\!\lceil\!\lceil [\mu, \underline{v}] \rceil\!\rceil\!\rceil_{\mathbb{K}\times V}$$

for all $\mu \in \mathbb{K}$, where the constant in this estimate depends only on the maximum number $\mathcal{Z}$ of elements with one common vertex.

An appropriate residual a posteriori error estimator can be defined by

$$\tilde{\eta}_T := \Big\{ h_{+,T}^2 \lceil\!\lceil \underline{R}_T \rceil\!\rceil_{0,T}^2 + \frac{1}{2} \sum_{E\in\mathcal{E}(T)\cap\mathcal{E}_{h,\Omega_S}} \frac{\lceil E \rceil}{\lceil T \rceil} h_{+,T}^2 \lceil\!\lceil \underline{R}_E \rceil\!\rceil_{0,E}^2 + \sum_{E\in\mathcal{E}(T)\cap\mathcal{E}_{h,N}} \frac{\lceil E \rceil}{\lceil T \rceil} h_{+,T}^2 \lceil\!\lceil \underline{R}_N \rceil\!\rceil_{0,E}^2 \Big\}^{1/2}. \tag{3.13}$$

Furthermore, let $\mathrm{R}_h = [0, \mathrm{I}_h] : \mathbb{K}\times V \to \mathbb{K}\times V_h$ be the interpolation operator on $\mathbb{K}\times V$ induced by I_h; that is, $\mathrm{R}_h[\mu, \underline{v}] = [0, \mathrm{I}_h\, \underline{v}]$. Combining the previous results, we obtain the following lemma and can immediately conclude an upper bound for the error $\|[\lambda, \underline{u}] - [\lambda_h, \underline{u}_h]\|_{\mathbb{K}\times V}$ from (2.11).

Lemma 3.15. *Let $\tilde{\eta}_T$ be the error estimator defined in (3.13). Then the estimate*

$$\left\|(\mathrm{Id}_{\mathbb{K}\times V} - \mathrm{R}_h)^{\#} F([\lambda_h, \underline{u}_h])\right\|_{\mathbb{K}\times V^\star} \lesssim \Big\{ \sum_{T\in\mathcal{T}_h} \tilde{\eta}_T^2 \Big\}^{1/2}$$

holds for all $\lambda_h \in \mathbb{K}$ and for all $\underline{u}_h \in V_h$ with $(\underline{u}_h, \bar{\underline{u}}_h)_H = 1$.

Proof. The assertion follows from

$$\|(\mathrm{Id}_{\mathbb{K}\times V} - \mathrm{R}_h)^{\#} F([\lambda_h, \underline{u}_h])\|_{\mathbb{K}\times V^\star} = \sup_{[\mu,\underline{v}]\in\mathbb{K}\times V} \frac{|\langle(\mathrm{Id}_{\mathbb{K}\times V} - \mathrm{R}_h)^{\#} F([\lambda_h, \underline{u}_h]), [\bar{\mu}, \bar{\underline{v}}]\rangle|}{\|[\mu, \underline{v}]\|_{\mathbb{K}\times V}}$$

and

$$\begin{aligned} |\langle(\mathrm{Id}_{\mathbb{K}\times V} - \mathrm{R}_h)^{\#} F([\lambda_h, \underline{u}_h]), [\bar{\mu}, \bar{\underline{v}}]\rangle| &= |\langle F([\lambda_h, \underline{u}_h]), (\mathrm{Id}_{\mathbb{K}\times V} - \mathrm{R}_h)[\bar{\mu}, \bar{\underline{v}}]\rangle| \\ = |\langle F([\lambda_h, \underline{u}_h]), [\bar{\mu}, \bar{\underline{v}} - \mathrm{I}_h\, \bar{\underline{v}}]\rangle| &\lesssim \Big\{ \sum_{T\in\mathcal{T}_h} \tilde{\eta}_T^2 \Big\}^{1/2} \cdot \|[\mu, \underline{v}]\|_{\mathbb{K}\times V}. \end{aligned}$$

□

Corollary 3.16. *Let $\lambda_0 \in \mathbb{K}$ be a simple eigenvalue of problem (3.2), and let $\underline{u}_0 \in V$ be a corresponding eigenfunction with $(\underline{u}_0, \bar{\underline{u}}_0)_H = 1$. Moreover, let $[\lambda_h, \underline{u}_h] \in \mathbb{K}\times V_h$ with $(\underline{u}_h, \bar{\underline{u}}_h)_H = 1$ be an eigenpair of the discretized eigenvalue problem (3.4) which approximates $[\lambda_0, \underline{u}_0]$ well enough so that the assumptions of Lemma 2.4 (page 16) are satisfied. Then, the estimate*

$$\|[\lambda_0, \underline{u}_0] - [\lambda_h, \underline{u}_h]\|_{\mathbb{K}\times V} \lesssim \Big\{ \sum_{T\in\mathcal{T}_h} \tilde{\eta}_T^2 \Big\}^{1/2}$$

holds.

Proof. Relation (2.11) and Lemma 2.5 imply that

$$\begin{aligned} \|[\lambda_0, \underline{u}_0] - [\lambda_h, \underline{u}_h]\|_{\mathbb{K}\times V} &\sim \|F([\lambda_h, \underline{u}_h])\|_{\mathbb{K}\times V^\star} \\ &\leq \|(\mathrm{Id}_{\mathbb{K}\times V} - \mathrm{R}_h)^{\#} F([\lambda_h, \underline{u}_h])\|_{\mathbb{K}\times V^\star} \\ &\quad + \|\mathrm{R}_h\|_{\mathcal{L}(\mathbb{K}\times V, \mathbb{K}\times V_h)} \|F([\lambda_h, \underline{u}_h]) - F_h([\lambda_h, \underline{u}_h])\|_{\mathbb{K}\times V_h^\star} \\ &\quad + \|\mathrm{R}_h\|_{\mathcal{L}(\mathbb{K}\times V, \mathbb{K}\times V_h)} \|F_h([\lambda_h, \underline{u}_h])\|_{\mathbb{K}\times V_h^\star}, \end{aligned}$$

Due to the definition of F_h (see page 37), the error $\|F([\lambda_h, \underline{u}_h]) - F_h([\lambda_h, \underline{u}_h])\|_{\mathbb{K}\times V_h^\star}$ vanishes. Moreover, $\|F_h([\lambda_h, \underline{u}_h])\|_{\mathbb{K}\times V_h^\star} = 0$ by assumption. The assertion follows from Lemma 3.15.

□

In order to obtain a lower bound for the error as well, we use residuals from a finite dimensional space, which is common practice, see standard textbooks, e.g. [104]. We replace the element and edge residuals $\underline{R}_T$, $\underline{R}_E$ and $\underline{R}_N$ by their L^2-projections

$$\begin{aligned}
\pi_{0,T}\underline{R}_T &:= \frac{1}{\lceil T \rceil}\int_T \underline{R}_T \,\mathrm{d}S \qquad \text{for } T \in \mathcal{T}_h,\\
\pi_{0,E}\underline{R}_E &:= \frac{1}{\lceil E \rceil}\int_E \underline{R}_E \,\mathrm{d}\sigma \qquad \text{for } E \in \mathcal{E}_{h,\Omega_S},\\
\pi_{0,E}\underline{R}_N &:= \frac{1}{\lceil E \rceil}\int_E \underline{R}_N \,\mathrm{d}\sigma \qquad \text{for } E \in \mathcal{E}_{h,N}.
\end{aligned} \tag{3.14}$$

Defining the residual a posteriori error estimator by

$$\begin{aligned}
\eta_T := \Big\{ h_{+,T}^2 \lVert \pi_{0,T}\underline{R}_T \rVert_{0,T}^2 &+ \frac{1}{2}\sum_{E\in\mathcal{E}(T)\cap\mathcal{E}_{h,\Omega_S}} \frac{\lceil E \rceil}{\lceil T \rceil} h_{+,T}^2 \lVert \pi_{0,E}\underline{R}_E \rVert_{0,E}^2 \\
&+ \sum_{E\in\mathcal{E}(T)\cap\mathcal{E}_{h,N}} \frac{\lceil E \rceil}{\lceil T \rceil} h_{+,T}^2 \lVert \pi_{0,E}\underline{R}_N \rVert_{0,E}^2 \Big\}^{1/2}
\end{aligned} \tag{3.15}$$

and collecting the approximation errors in

$$\begin{aligned}
\varepsilon_T := \Big\{ h_{+,T}^2 \lVert \underline{R}_T - \pi_{0,T}\underline{R}_T \rVert_{0,T}^2 &+ \frac{1}{2}\sum_{E\in\mathcal{E}(T)\cap\mathcal{E}_{h,\Omega_S}} \frac{\lceil E \rceil}{\lceil T \rceil} h_{+,T}^2 \lVert \underline{R}_E - \pi_{0,E}\underline{R}_E \rVert_{0,E}^2 \\
&+ \sum_{E\in\mathcal{E}(T)\cap\mathcal{E}_{h,N}} \frac{\lceil E \rceil}{\lceil T \rceil} h_{+,T}^2 \lVert \underline{R}_N - \pi_{0,E}\underline{R}_N \rVert_{0,E}^2 \Big\}^{1/2},
\end{aligned} \tag{3.16}$$

we obtain the following upper error bound.

Theorem 3.17 (Upper error bound). *Let $\lambda_0 \in \mathbb{K}$ be a simple eigenvalue of problem (3.2), and let $\underline{u}_0 \in V$ be a corresponding eigenfunction with $(\underline{u}_0, \bar{\underline{u}}_0)_H = 1$. Moreover, let $[\lambda_h, \underline{u}_h] \in \mathbb{K} \times V_h$ with $(\underline{u}_h, \bar{\underline{u}}_h)_H = 1$ be an eigenpair of the discretized eigenvalue problem (3.4) which approximates $[\lambda_0, \underline{u}_0]$ well enough so that the assumptions of Lemma 2.4 are satisfied. Then the estimate*

$$|||[\lambda_0, \underline{u}_0] - [\lambda_h, \underline{u}_h]|||_{\mathbb{K}\times V} \lesssim \Big\{ \sum_{T\in\mathcal{T}_h} \eta_T^2 \Big\}^{1/2} + \Big\{ \sum_{T\in\mathcal{T}_h} \varepsilon_T^2 \Big\}^{1/2}$$

holds.

Proof. The assertion follows from the relations $\tilde{\eta}_T^2 \leq \eta_T^2 + \varepsilon_T^2$ and $\{\sum_{T\in\mathcal{T}_h}(\eta_T^2 + \varepsilon_T^2)\}^{1/2} \leq \{\sum_{T\in\mathcal{T}_h} \eta_T^2\}^{1/2} + \{\sum_{T\in\mathcal{T}_h} \varepsilon_T^2\}^{1/2}$ and from Corollary 3.16. □

Remark 3.18. *The condition that $[\lambda_h, \underline{u}_h]$ is close enough to $[\lambda_0, \underline{u}_0]$ means, in particular, that λ_h is closer to λ_0 than to any other eigenvalue of problem (3.2) and that the assumptions of Lemma 2.10 are satisfied, cf. Remark 2.11.*

The constant in the estimate of Theorem 3.17 depends on the constants in the relation (2.11) and Theorem 3.11 and on the maximum number of elements with a common vertex, see also Section 3.7.

According to Remark 2.9, the condition $(\underline{u}_0, \bar{\underline{u}}_0)_H \neq 0$ has an influence only on the constants in the upper error bounds in Corollary 3.16 and Theorem 3.17. The specific value of the term $(\underline{u}, \bar{\underline{u}})_H$ is irrelevant for the constants. The suggested value 1 can be replaced by any other complex number $z \neq 0$, if the function F is redefined appropriately. The effect is that both side in the error estimates are scaled by the same factor. When $(\underline{u}, \bar{\underline{u}})_H = 0$, nothing can be said about the validity of the upper error bound. The lower error bound (which will be stated in Theorem 3.23) holds nevertheless.

3.5 A lower error bound

In the previous section, we proved the reliability of the residual a posteriori error estimator defined in (3.15). This section is aimed at proving its efficiency, that is, we will show that η_T provides a lower bound for the error of the solutions to the eigenvalue problem (3.2). To this end, let $\mathfrak{b}_T$ and $\mathfrak{b}_E$ be some element and edge bubble functions corresponding to $T \in \mathcal{T}_h$ and $E \in \mathcal{T}_h$, respectively, so that the following properties are satisfied:

$$\begin{aligned} &\operatorname{supp} \mathfrak{b}_T \subseteq \overline{T}, && 0 \le \mathfrak{b}_T(\varphi,\theta) \le 1 \quad \forall(\varphi,\theta) \in T, && \max_{(\varphi,\theta)\in T} \mathfrak{b}_T(\varphi,\theta) = 1 \\ &\operatorname{supp} \mathfrak{b}_E \subseteq \overline{\omega}_E, && 0 \le \mathfrak{b}_E(\varphi,\theta) \le 1 \quad \forall(\varphi,\theta) \in \omega_E, && \max_{(\varphi,\theta)\in E} \mathfrak{b}_E(\varphi,\theta) = 1, \end{aligned} \tag{3.17}$$

$$\begin{aligned} \int_T \mathfrak{b}_T \,\mathrm{d}\mathcal{S} &\sim \lceil T \rceil \sim \int_T \mathfrak{b}_T^2 \,\mathrm{d}\mathcal{S}, \\ \int_E \mathfrak{b}_E \,\mathrm{d}\sigma &\sim \lceil E \rceil \sim \int_E \mathfrak{b}_E^2 \,\mathrm{d}\sigma, \\ \int_{T'} \mathfrak{b}_E \,\mathrm{d}\mathcal{S} &\sim \lceil T' \rceil \sim \int_{T'} \mathfrak{b}_E^2 \,\mathrm{d}\mathcal{S} \quad \text{for all } T' \subset \omega_E, \end{aligned} \tag{3.18}$$

$$\|\lceil \nabla_{\mathcal{S}} \mathfrak{b}_T \|_{0,T} \lesssim h_{-,T'}^{-1} \lceil T \rceil \tag{3.19}$$

$$\|\lceil \nabla_{\mathcal{S}} \mathfrak{b}_E \|_{0,T'} \lesssim h_{-,T'}^{-1} \lceil T \rceil \quad \text{for all } T' \subset \omega_E. \tag{3.20}$$

Remark 3.19. *The bubble functions can be defined, for example, via the nodal basis functions $\phi_{\mathrm{x}_{i,T}}$, $i = 1, \ldots, 3$, of an element $T \in \mathcal{T}_h$. For $T \in \mathcal{T}_h$, the element bubble function is given by*

$$\mathfrak{b}_T := \begin{cases} 27 \phi_{\mathrm{x}_{1,T}} \phi_{\mathrm{x}_{2,T}} \phi_{\mathrm{x}_{2,T}} & \text{on } T \\ 0 & \text{on } \Omega_S \setminus T. \end{cases}$$

For an edge $E \in \mathcal{E}_{h,\Omega}$ with $\omega_E = T_1 \cup T_2$, $T_1, T_2 \in \mathcal{T}_h$, let the nodes of T_1 and T_2 be enumerated so that the nodes of E are counted first. The definition of the corresponding edge-bubble function

$$\mathfrak{b}_E := \begin{cases} 4 \phi_{\mathrm{x}_{1,T_i}} \phi_{\mathrm{x}_{2,T_i}} & \text{on } T_i,\ i = 1, 2 \\ 0 & \text{on } \Omega_S \setminus \omega_E \end{cases}$$

can easily be extended to boundary edges, where $T_1 = T_2 = \omega_E$.

If the bubble functions are defined in the proposed way, the properties (3.17)–(3.20) follow in the Cartesian case (that is for polygons or polyhedra) from standard scaling arguments. For domains on the sphere, however, the verification of these estimates requires a more careful analysis, since the weighted surface and line elements do not allow for the same trivial conclusions. Examples for the nodal basis functions in spherical coordinates, which provide bubble functions with the desired properties, were given in Remark 3.10.

We introduce the finite dimensional space

$$\tilde{V}_h := \left[\operatorname{span} \{ \mathfrak{b}_T, \mathfrak{b}_E \mid T \in \mathcal{T}_h,\ E \in \mathcal{E}_{h,\Omega_S} \cup \mathcal{E}_{h,N} \} \right]^d$$

which consists of linear combinations of functions that vanish on ∂T for $T \in \mathcal{T}_h$ or on $\partial \omega_E$ for $E \in \mathcal{E}_{h,\Omega_S} \cup \mathcal{E}_{h,N}$. Moreover, we define the approximation $\tilde{F}_h : \mathbb{K} \times V \to \mathbb{K} \times \tilde{V}_h^\star$,

$$\begin{aligned} \langle \tilde{F}_h([\lambda_h, \underline{u}_h]), [\bar{\mu}, \underline{\bar{v}}] \rangle := &\sum_{T \in \mathcal{T}_h} \int_T \pi_{0,T} \underline{R}_T \cdot \underline{v} \,\mathrm{d}\mathcal{S} - \sum_{E \in \mathcal{E}_{h,\Omega_S}} \int_E \pi_{0,E} \underline{R}_E \cdot \underline{v} \,\mathrm{d}\sigma \\ &+ \sum_{E \in \mathcal{E}_{h,N}} \int_E \pi_{0,E} \underline{R}_N \cdot \underline{v} \,\mathrm{d}\sigma, \end{aligned} \tag{3.21}$$

of the operator $F : \mathbb{K} \times V \to \mathbb{K} \times V^\star$. We will prove that

$$\eta_T \lesssim \sup_{\substack{\underline{v} \in \tilde{V}_h \\ \operatorname{supp} \underline{v} \subseteq \overline{\tilde{\omega}}_T}} \frac{|\langle \tilde{F}_h([\lambda_h, \underline{u}_h]), [0, \bar{\underline{v}}]\rangle|}{\|\underline{v}\|_{1,T}} \qquad \forall T \in \mathcal{T}_h. \tag{3.22}$$

The argumentation is similar to the Cartesian case, which was outlined, for example, by Verfürth [104]. Nevertheless, the constants in the estimates differ due to the spherical nature of the problem and the resulting differences in the definition of the norms, integrals and nodal basis functions. For completeness and in order to give a coherent overview of all the necessary ingredients, the entire proof will be presented on the next pages.

For $T \in \mathcal{T}_h$, we define $\underline{w}_T := (\overline{\pi_{0,T}\underline{R}_T})\, \mathfrak{b}_T \in \tilde{V}_h$. This function vanishes on ∂T and satisfies $\operatorname{supp} \underline{w}_T \subseteq \overline{T}$ and therefore $\nabla_{\mathcal{S}} \underline{w}_T \equiv 0$ on $\Omega_{\mathcal{S}} \setminus T$. The relations (3.18) and (3.21) imply that

$$\begin{aligned} \|\pi_{0,T}\underline{R}_T\|_{0,T}^2 &= |\pi_{0,T}\underline{R}_T|^2 \lceil T \rceil \sim |\pi_{0,T}\underline{R}_T|^2 \int_T \mathfrak{b}_T \, \mathrm{d}\mathcal{S} \\ &= \int_T (\pi_{0,T}\underline{R}_T) \cdot \underline{w}_T \, \mathrm{d}\mathcal{S} = \langle \tilde{F}_h([\lambda_h, \underline{u}_h]), [0, \bar{\underline{w}}_T]\rangle. \end{aligned} \tag{3.23}$$

For $E \in \mathcal{E}_{h,\Omega_{\mathcal{S}}}$, we define $\underline{w}_E := (\overline{\pi_{0,E}\underline{R}_E})\, \mathfrak{b}_E \in \tilde{V}_h$. This function vanishes on $\partial\omega_E$ and satisfies $\operatorname{supp} \overline{\underline{w}_E} \subseteq \omega_E$ and therefore $\nabla_{\mathcal{S}} \underline{w}_E \equiv 0$ on $\Omega_{\mathcal{S}} \setminus \omega_E$. The relations (3.18) and (3.21) imply that

$$\begin{aligned} \|\pi_{0,E}\underline{R}_E\|_{0,E}^2 &= |\pi_{0,E}\underline{R}_E|^2 \lceil E \rceil \sim |\pi_{0,E}\underline{R}_E|^2 \int_E \mathfrak{b}_E \, \mathrm{d}\sigma \\ &= \int_E (\pi_{0,E}\underline{R}_E) \cdot \underline{w}_E \, \mathrm{d}\sigma \\ &= -\langle \tilde{F}_h([\lambda_h, \underline{u}_h]), [0, \bar{\underline{w}}_E]\rangle + \sum_{T \subset \omega_E} \int_T (\pi_{0,T}\underline{R}_T) \cdot \underline{w}_E \, \mathrm{d}\mathcal{S}. \end{aligned} \tag{3.24}$$

Lemma 3.20. *The functions $\underline{w}_T$ and $\underline{w}_E$ satisfy*

$$\begin{aligned} \|\underline{w}_T\|_{1,T} &\lesssim h_{-,T}^{-1} \|\pi_{0,T}\underline{R}_T\|_{0,T}, \\ \|\underline{w}_E\|_{0,\omega_E} &\lesssim \|\pi_{0,E}\underline{R}_E\|_{0,E} \sum_{T \subset \omega_E} \frac{\lceil T \rceil^{1/2}}{\lceil E \rceil^{1/2}}, \\ \|\underline{w}_E\|_{1,\omega_E} &\lesssim \|\pi_{0,E}\underline{R}_E\|_{0,E} \sum_{T \subset \omega_E} h_{-,T}^{-1} \frac{\lceil T \rceil^{1/2}}{\lceil E \rceil^{1/2}}. \end{aligned}$$

Proof. The relations (3.18)–(3.20) imply that

$$\begin{aligned} \|\underline{w}_T\|_{1,T}^2 &= \|(\overline{\pi_{0,T}\underline{R}_T})\, \mathfrak{b}_T\|_{1,T}^2 = |\pi_{0,T}\underline{R}_T|^2 \|\mathfrak{b}_T\|_{1,T}^2 \\ &= |\pi_{0,T}\underline{R}_T|^2 \left(\frac{1}{4}\|\mathfrak{b}_T\|_{0,T}^2 + \|\nabla_{\mathcal{S}}\mathfrak{b}_T\|_{0,T}^2\right) \\ &\lesssim |\pi_{0,T}\underline{R}_T|^2 \left(\lceil T \rceil + h_{-,T}^{-2} \lceil T \rceil\right) \\ &\lesssim h_{-,T}^{-2} \|\pi_{0,T}\underline{R}_T\|_{0,T}^2 \end{aligned}$$

for $T \in \mathcal{T}_h$ and

$$
\begin{aligned}
\lceil \underline{w}_E \rceil_{0,\omega_E}^2 &= \lceil (\overline{\pi_{0,E}\underline{R}_E})\, \mathfrak{b}_E \rceil_{0,\omega_E}^2 = |\pi_{0,E}\underline{R}_E|^2 \lceil \mathfrak{b}_E \rceil_{0,\omega_E}^2 \\
&\sim |\pi_{0,E}\underline{R}_E|^2 \sum_{T\subset\omega_E} \lceil T \rceil = \lceil \pi_{0,E}\underline{R}_E \rceil_{0,E}^2 \sum_{T\subset\omega_E} \frac{\lceil T \rceil}{\lceil E \rceil}, \\
\lceil \underline{w}_E \rceil_{1,\omega_E}^2 &= \lceil (\overline{\pi_{0,E}\underline{R}_E})\, \mathfrak{b}_E \rceil_{1,\omega_E}^2 = |\pi_{0,E}\underline{R}_E|^2 \lceil \mathfrak{b}_E \rceil_{1,\omega_E}^2 \\
&= |\pi_{0,E}\underline{R}_E|^2 \sum_{T\subset\omega_E} \left(\frac{1}{4}\lceil \mathfrak{b}_E \rceil_{0,T}^2 + \lceil \nabla_S \mathfrak{b}_E \rceil_{0,T}^2\right) \\
&\lesssim |\pi_{0,E}\underline{R}_E|^2 \sum_{T\subset\omega_E} \left(\lceil T \rceil + h_{-,T}^{-2} \lceil T \rceil\right) \\
&\lesssim \lceil \pi_{0,E}\underline{R}_E \rceil_{0,E}^2 \sum_{T\subset\omega_E} h_{-,T}^{-2} \frac{\lceil T \rceil}{\lceil E \rceil}
\end{aligned}
$$

for $E \in \mathcal{E}_h$. □

Collecting the previous estimates, we conclude from Lemma 3.20 and the relations (3.23) and (3.24) that

$$
\begin{aligned}
\lceil \pi_{0,T}\underline{R}_T \rceil_{0,T}^2 &\lesssim \lceil \underline{w}_T \rceil_{1,T}^{-1} h_{-,T}^{-1} \lceil \pi_{0,T}\underline{R}_T \rceil_{0,T} \langle \tilde{F}_h([\lambda_h, \underline{u}_h]), [0, \bar{\underline{w}}_T] \rangle, \\
\lceil \pi_{0,E}\underline{R}_E \rceil_{0,E}^2 &\lesssim \lceil \underline{w}_E \rceil_{1,\omega_E}^{-1} \sum_{T\subset\omega_E} h_{-,T}^{-1} \frac{\lceil T \rceil^{1/2}}{\lceil E \rceil^{1/2}} \lceil \pi_{0,E}\underline{R}_E \rceil_{0,E} |\langle \tilde{F}_h([\lambda_h, \underline{u}_h]), [0, \bar{\underline{w}}_E] \rangle| \\
&\quad + \sum_{T\subset\omega_E} \lceil \pi_{0,T}\underline{R}_T \rceil_{0,T} \lceil \underline{w}_E \rceil_{0,T} \\
&\lesssim \lceil \pi_{0,E}\underline{R}_E \rceil_{0,E} \Big\{ \sum_{T\subset\omega_E} h_{-,T}^{-1} \frac{\lceil T \rceil^{1/2}}{\lceil E \rceil^{1/2}} \lceil \underline{w}_E \rceil_{1,\omega_E}^{-1} |\langle \tilde{F}_h([\lambda_h, \underline{u}_h]), [0, \bar{\underline{w}}_E] \rangle| \\
&\quad + \sum_{T\subset\omega_E} \frac{\lceil T \rceil^{1/2}}{\lceil E \rceil^{1/2}} \lceil \pi_{0,T}\underline{R}_T \rceil_{0,T} \Big\}.
\end{aligned}
$$

In both estimates, we divide by the residual norm. Inserting the first inequality into the second one, we obtain that

$$
\begin{aligned}
\lceil \pi_{0,E}\underline{R}_E \rceil_{0,E} &\lesssim \sum_{T\subset\omega_E} h_{-,T}^{-1} \frac{\lceil T \rceil^{1/2}}{\lceil E \rceil^{1/2}} \lceil \underline{w}_E \rceil_{1,\omega_E}^{-1} |\langle \tilde{F}_h([\lambda_h, \underline{u}_h]), [0, \bar{\underline{w}}_E] \rangle| \\
&\quad + \sum_{T\subset\omega_E} h_{-,T}^{-1} \frac{\lceil T \rceil^{1/2}}{\lceil E \rceil^{1/2}} \lceil \underline{w}_T \rceil_{1,T}^{-1} |\langle \tilde{F}_h([\lambda_h, \underline{u}_h]), [0, \bar{\underline{w}}_T] \rangle| \\
&\lesssim \sum_{T\subset\omega_E} h_{-,T}^{-1} \frac{\lceil T \rceil^{1/2}}{\lceil E \rceil^{1/2}} \sup_{\substack{\underline{v} \in \tilde{V}_h \\ \operatorname{supp} \underline{v} \subseteq \bar{\omega}_E}} \frac{|\langle \tilde{F}_h([\lambda_h, \underline{u}_h]), [0, \bar{\underline{v}}] \rangle|}{\lceil \underline{v} \rceil_{1,\omega_E}}.
\end{aligned}
$$

Since the size of adjacent elements does not change rapidly, we conclude that

$$
\lceil \pi_{0,E}\underline{R}_E \rceil_{0,E} \lesssim h_{-,T}^{-1} \frac{\lceil T \rceil^{1/2}}{\lceil E \rceil^{1/2}} \sup_{\substack{\underline{v} \in \tilde{V}_h \\ \operatorname{supp} \underline{v} \subseteq \bar{\omega}_E}} \frac{|\langle \tilde{F}_h([\lambda_h, \underline{u}_h]), [0, \bar{\underline{v}}] \rangle|}{\lceil \underline{v} \rceil_{1,\omega_E}}
$$

for each $T \subset \omega_E$. Hence,

$$h_{-,T} \| \pi_{0,T} \underline{R}_T \|_{0,T} \lesssim \sup_{\substack{\underline{v} \in \tilde{V}_h \\ \operatorname{supp} \underline{v} \subseteq \overline{T}}} \frac{|\langle \tilde{F}_h([\lambda_h, \underline{u}_h]), [0, \underline{v}] \rangle|}{\| \underline{v} \|_{1,T}}, \tag{3.25}$$

$$\frac{\lceil E \rceil^{1/2}}{\lceil T \rceil^{1/2}} h_{-,T} \| \pi_{0,E} \underline{R}_E \|_{0,E} \lesssim \sup_{\substack{\underline{v} \in \tilde{V}_h \\ \operatorname{supp} \underline{v} \subseteq \overline{\omega}_E}} \frac{|\langle \tilde{F}_h([\lambda_h, \underline{u}_h]), [0, \underline{v}] \rangle|}{\| \underline{v} \|_{1,\omega_E}} \tag{3.26}$$

for all $T \in \mathcal{T}_h$ and for all $E \in \mathcal{E}(T)$, respectively.

For edges $E \in \mathcal{E}_{h,N}$ at the Neumann boundary, we define $\underline{w}_E := (\overline{\pi_{0,E} \underline{R}_N}) \mathfrak{b}_E$. The patch ω_E consists of only one element in this case. Nevertheless, the same arguments as above can be used to prove that

$$\frac{\lceil E \rceil^{1/2}}{\lceil T \rceil^{1/2}} h_{-,T} \| \pi_{0,T} \underline{R}_N \|_{0,E} \lesssim \sup_{\substack{\underline{v} \in \tilde{V}_h \\ \operatorname{supp} \underline{v} \subseteq \overline{\omega}_E}} \frac{|\langle \tilde{F}_h([\lambda_h, \underline{u}_h]), [0, v] \rangle|}{\| \underline{v} \|_{1,T}} \tag{3.27}$$

for all $E \in \mathcal{E}_{h,N}$.

To prove (3.22), we have to employ the isotropy condition (3.6) which entails

$$h_{+,T} \lesssim h_{-,T}.$$

The combination of the estimates (3.25)–(3.27) with the definition of the residual error estimator (3.15) on page 49 yields that

$$\begin{aligned} \eta_T &\leq h_{+,T} \Big(\| \pi_{0,T} \underline{R}_T \|_{0,T} + \frac{1}{2} \sum_{E \in \mathcal{E}(T) \cap \mathcal{E}_{h,\Omega_S}} \frac{\lceil E \rceil^{1/2}}{\lceil T \rceil^{1/2}} \| \pi_{0,E} \underline{R}_E \|_{0,E} \\ &\qquad + \sum_{E \in \mathcal{E}(T) \cap \mathcal{E}_{h,N}} \frac{\lceil E \rceil^{1/2}}{\lceil T \rceil^{1/2}} \| \pi_{0,T} \underline{R}_N \|_{0,E} \Big) \\ &\lesssim h_{-,T} \Big(\| \pi_{0,T} \underline{R}_T \|_{0,T} + \frac{1}{2} \sum_{E \in \mathcal{E}(T) \cap \mathcal{E}_{h,\Omega_S}} \frac{\lceil E \rceil^{1/2}}{\lceil T \rceil^{1/2}} \| \pi_{0,E} \underline{R}_E \|_{0,E} \\ &\qquad + \sum_{E \in \mathcal{E}(T) \cap \mathcal{E}_{h,N}} \frac{\lceil E \rceil^{1/2}}{\lceil T \rceil^{1/2}} \| \pi_{0,T} \underline{R}_N \|_{0,E} \Big) \\ &\lesssim \sup_{\substack{\underline{v} \in \tilde{V}_h \\ \operatorname{supp} \underline{v} \subseteq \overline{\omega}_T}} \frac{|\langle \tilde{F}_h([\lambda_h, \underline{u}_h]), [0, \underline{v}] \rangle|}{\| \underline{v} \|_{1,\omega_T}}. \end{aligned}$$

As in the previous section, we exploit that the number of patches to which an element T belongs is bounded from above by a constant which is independent of the discretization parameter h. Due to (3.1), we obtain the following lemma by summing up over all elements $T \in \mathcal{T}_h$.

Lemma 3.21. *The relation*

$$\Big\{ \sum_{T \in \mathcal{T}_h} \eta_T^2 \Big\}^{1/2} \lesssim \| \tilde{F}_h([\lambda_h, \underline{u}_h]) \|_{\mathbb{K} \times \tilde{V}_h^\star}$$

holds for all $[\lambda_h, \underline{u}_h] \in \mathbb{K} \times V_h$ *with* $(\underline{u}_h, \overline{\underline{u}}_h)_H = 1$.

Lemma 3.22. *The relation*

$$\|F([\lambda_h, \underline{u}_h]) - \tilde{F}_h([\lambda_h, \underline{u}_h])\|_{\mathbb{K}\times\tilde{V}_h^*} \lesssim \Big\{ \sum_{T\in\mathcal{T}_h} \varepsilon_T^2 \Big\}^{1/2}$$

holds for all $[\lambda_h, \underline{u}_h] \in \mathbb{K} \times V_h$ *with* $(\underline{u}_h, \bar{\underline{u}}_h)_H = 1$.

Proof. By definition,

$$\|F([\lambda_h, \underline{u}_h]) - \tilde{F}_h([\lambda_h, \underline{u}_h])\|_{\mathbb{K}\times\tilde{V}_h^*} = \sup_{[\mu,\underline{v}]\,\in\,\mathbb{K}\,\times\,\tilde{V}_h} \frac{|\langle F([\lambda_h, \underline{u}_h]) - \tilde{F}_h([\lambda_h, \underline{u}_h]), [\bar{\mu}, \bar{\underline{v}}]\rangle|}{\|[\mu, \underline{v}]\|_{\mathbb{K}\times V}}.$$

Let $T \in \mathcal{T}_h$ be an arbitrary element. Each $\underline{v} \in \tilde{V}_h$ vanishes on at least one edge of T. The Friedrichs-type inequality (Lemma 3.14) and the trace theorem (Lemma 3.12) yield that

$$\|\underline{v}\|_{0,T} \lesssim h_{+,T} |\underline{v}|_{1,T}, \qquad \|\underline{v}\|_{0,E} \lesssim \frac{|E|^{1/2}}{|T|^{1/2}} h_{+,T} |\underline{v}|_{1,T}.$$

Hence,

$$\begin{aligned}
&|\langle F([\lambda_h, \underline{u}_h]) - \tilde{F}_h([\lambda_h, \underline{u}_h]), [\bar{\mu}, \bar{\underline{v}}]\rangle| \\
&\quad\le \sum_{T\in\mathcal{T}_h} \Big\{ \|\underline{R}_T - \pi_{0,T}\underline{R}_T\|_{0,T} \|\underline{v}\|_{0,T} + \frac{1}{2} \sum_{E\in\mathcal{E}(T)\cap\mathcal{E}_{h,\Omega_S}} \|\underline{R}_E - \pi_{0,E}\underline{R}_E\|_{0,E} \|\underline{v}\|_{0,E} \\
&\qquad\qquad + \sum_{E\in\mathcal{E}(T)\cap\mathcal{E}_{h,N}} \|\underline{R}_N - \pi_{0,E}\underline{R}_N\|_{0,E} \|\underline{v}\|_{0,E} \Big\} \\
&\quad\lesssim \Big\{ \sum_{T\in\mathcal{T}_h} \varepsilon_T^2 \Big\}^{1/2} \cdot \Big\{ \sum_{T\in\mathcal{T}_h} \Big(|\underline{v}|_{1,T}^2 + \frac{1}{2} \sum_{E\in\mathcal{E}(T)\cap\mathcal{E}_{h,\Omega_S}} |\underline{v}|_{1,T}^2 + \sum_{E\in\mathcal{E}(T)\cap\mathcal{E}_{h,N}} |\underline{v}|_{1,T}^2 \Big) \Big\}^{1/2} \\
&\quad\lesssim \Big\{ \sum_{T\in\mathcal{T}_h} \varepsilon_T^2 \Big\}^{1/2} \cdot |\underline{v}|_{1,\Omega_S}^2 \lesssim \Big\{ \sum_{T\in\mathcal{T}_h} \varepsilon_T^2 \Big\}^{1/2} \cdot \|[\mu, \underline{v}]\|_{\mathbb{K}\times V}
\end{aligned}$$

for all $\mu \in \mathbb{K}$, $\underline{v} \in V$. This completes the proof. □

The combination of Lemma 3.21, Lemma 3.22 and the relation (2.11) provides a lower error bound.

Theorem 3.23 (Lower error bound). *Let* $\lambda_0 \in \mathbb{K}$ *be a simple eigenvalue of problem (3.2), and let* $\underline{u}_0 \in V$ *be a corresponding eigenfunction with* $(\underline{u}_0, \bar{\underline{u}}_0)_H = 1$. *Moreover, let* $[\lambda_h, \underline{u}_h] \in \mathbb{K}\times V_h$ *with* $(\underline{u}_h, \bar{\underline{u}}_h)_H = 1$ *be an eigenpair of the discretized eigenvalue problem (3.4) which approximates* $[\lambda_0, \underline{u}_0]$ *well enough so that the assumptions of Lemma 2.4 (page 16) are satisfied. Then the estimate*

$$\Big\{ \sum_{T\in\mathcal{T}_h} \eta_T^2 \Big\}^{1/2} \lesssim \|[\lambda, \underline{u}] - [\lambda_h, \underline{u}_h]\|_{\mathbb{K}\times V} + \Big\{ \sum_{T\in\mathcal{T}_h} \varepsilon_T^2 \Big\}^{1/2}$$

holds, where η_T *and* ε_T *are defined in (3.15) and (3.16).*

Proof. Note that $\|F([\lambda_h, \underline{u}_h])\|_{\mathbb{K}\times\tilde{V}_h^\star} \leq \|F([\lambda_h, \underline{u}_h])\|_{\mathbb{K}\times V^\star}$ since $\tilde{V}_h \subset V$. Lemma 3.21, the triangle inequality, Lemma 2.4 relation (2.11) and Lemma 3.22 imply that

$$\begin{aligned}
\Big\{\sum_{T\in\mathcal{T}_h}\eta_T^2\Big\}^{1/2} &\lesssim \|F([\lambda_h, \underline{u}_h])\|_{\mathbb{K}\times\tilde{V}_h^\star} + \|F([\lambda_h, \underline{u}_h]) - \tilde{F}_h([\lambda_h, \underline{u}_h])\|_{\mathbb{K}\times\tilde{V}_h^\star} \\
&\leq \|F([\lambda_h, \underline{u}_h])\|_{\mathbb{K}\times V^\star} + \|F([\lambda_h, \underline{u}_h]) - \tilde{F}_h([\lambda_h, \underline{u}_h])\|_{\mathbb{K}\times\tilde{V}_h^\star} \\
&\lesssim \|[\lambda_0, \underline{u}_0] - [\lambda_h, \underline{u}_h]\|_{\mathbb{K}\times V} + \Big\{\sum_{T\in\mathcal{T}_h}\varepsilon_T^2\Big\}^{1/2}.
\end{aligned}$$

□

Theorem 3.23 yields the efficiency of the error estimator η_T upto an addend of approximation errors ε_T. The following lemma shows that the terms ε_T are of higher order and can be neglected.

Lemma 3.24. *The approximation errors ε_T satisfy*

$$\varepsilon_T \lesssim h_{+,T}^2\Big\{\lceil \underline{R}_T\rceil_{1,T}^2 + \frac{1}{2}\sum_{E\in\mathcal{E}(T)\cap\mathcal{E}_{h,\Omega_S}}\frac{\lceil E\rceil^2}{\lceil T\rceil^2}\lceil \underline{R}_E\rceil_{1,T}^2 + \sum_{E\in\mathcal{E}(T)\cap\mathcal{E}_{h,N}}\frac{\lceil E\rceil^2}{\lceil T\rceil^2}\lceil \underline{R}_N\rceil_{1,T}^2\Big\}^{1/2}.$$

Proof. Let $T\in\mathcal{T}$ and $E\in\mathcal{E}(T)\cap(\mathcal{E}_{h,\Omega_S}\cup\mathcal{E}_{h,N})$. By definition, the terms $\pi_{0,T}\underline{R}_T$, $\pi_{0,E}\underline{R}_E$ and $\pi_{0,E}\underline{R}_N$ are constant over T. With the same argumentation as in the proof of the Poincaré-type inequality (Lemma 3.13), one can show that

$$\|\underline{R}_T - \pi_{0,T}\underline{R}_T\|_{0,T} \lesssim h_{+,T}\lceil \underline{R}_T\rceil_{1,T}, \qquad \|\underline{R}_E - \pi_{0,E}\underline{R}_E\|_{0,T} \lesssim h_{+,T}\lceil \underline{R}_E\rceil_{1,T}.$$

Note that all derivatives of $\pi_{0,E}\underline{R}_E$ vanish and thus $\lceil \underline{R}_E - \pi_{0,E}\underline{R}_E\rceil_{1,T} = \lceil \underline{R}_E\rceil_{1,T}$. From the trace theorem (Lemma 3.12), we conclude that

$$\begin{aligned}
\|\underline{R}_E - \pi_{0,E}\underline{R}_E\|_{0,E} &\lesssim \frac{\lceil E\rceil^{1/2}}{\lceil T\rceil^{1/2}}\Big(\|\underline{R}_E - \pi_{0,E}\underline{R}_E\|_{0,T}^2 + h_{+,T}^2\lceil \underline{R}_E - \pi_{0,E}\underline{R}_E\rceil_{1,T}^2\Big)^{1/2} \\
&\lesssim \frac{\lceil E\rceil^{1/2}}{\lceil T\rceil^{1/2}}h_{+,T}\lceil \underline{R}_E\rceil_{1,T}.
\end{aligned}$$

Likewise, for $E\in\mathcal{E}(T)\cap\mathcal{E}_{h,N}$, we obtain that

$$\|\underline{R}_N - \pi_{0,E}\underline{R}_N\|_{0,E} \lesssim \lceil E\rceil^{1/2}/\lceil T\rceil^{1/2}h_{+,T}\lceil \underline{R}_N\rceil_{1,T}.$$

The combination with the definition of ε_T in (3.16) yields the assertion. □

Remark 3.25. *The constant in Theorem 3.23 depends on the maximum number of elements with one common vertex, on the constants in the estimates (2.11) and (3.17)–(3.20), and on the constants in Lemmas 3.14 and 3.12.*

The condition that $[\lambda_h, \underline{u}_h]$ approximates $[\lambda_0, \underline{u}_0]$ well enough means, in particular, that λ_h is closer to λ_0 than to any other eigenvalue of (3.2) and that the operator DF *is Lipschitz continuous, compare Lemma 2.10 and Remark 2.11.*

Concerning the lower error bound, the assumptions of Lemma 2.4 can be weakened; provided that $\mathrm{DF}([\lambda_0, \underline{u}_0])$ *does not vanish identically, it is not necessary to assume that $[\lambda_0, \underline{u}_0]$ is a* regular *solution to (3.2). This means that the estimate of Theorem 3.23 holds even if* λ_0 *is not a simple eigenvalue. Moreover, it is not necessary that* $(\underline{u}_0, \bar{\underline{u}}_0)_H \neq 0$*, compare Remarks 2.9 and 3.18. In Theorem 3.23, this condition is required only to match the definition of the function* F*. The condition can be omitted, when* F *is defined as the function* $F : \mathbb{K}\times V \to V^\star$*,* $\langle F([\lambda, \underline{u}]), \underline{v}\rangle = (\mathcal{B}(\lambda)\underline{u}, \underline{v})_V$*, compare (2.15).*

Remark 3.26. *Since we use constant approximations of the residuals, the functions* $\underline{w}_E = (\overline{\pi_{0,E}\underline{R}_E})\, \mathfrak{b}_E$ *and* $\underline{w}_E = (\overline{\pi_{0,E}\underline{R}_N})\, \mathfrak{b}_E$ *are defined on* ω_E *without the need of an extension (or prolongation) operator. The resulting addends* $\{\sum_{T\in\mathcal{T}_h} \varepsilon_T^2\}^{1/2}$ *in the final error bounds in Theorems 3.17 and 3.23 are higher-order terms according to Lemma 3.24.*

If the residuals $\underline{R}_T$*,* $\underline{R}_E$ *and* $\underline{R}_N$ *live in a finite dimensional space, one can show (with much heavier machinery) that*

$$\Big\{\sum_{T\in\mathcal{T}_h} \tilde{\eta}_T^2\Big\}^{1/2} \lesssim \|[\lambda, \underline{u}] - [\lambda_h, \underline{u}_h]\|_{\mathbb{K}\times V}, \tag{3.28}$$

where $\tilde{\eta}_T$ *is defined in (3.13) on page 47. Note that, owing to the spherical nature of the eigenvalue problem, the (unprojected) residuals are usually not linear combinations of polynomials. To prove (3.28), we need a larger finite-dimensional space* $\tilde{Y}_h$ *and an extension operator* $\mathrm{F}_{\mathrm{ext}}$ *which extends the domain of a function* $w \in [\mathcal{H}^1(E)]^d$ *from an edge* E *to the patch* ω_E *so that* $\mathrm{F}_{\mathrm{ext}}(w) \in [\mathcal{H}^1(\omega_E)]^d$*. The estimates (3.23), (3.24) as well as the assertions of Lemma 3.20 are verified with arguments of equivalent norms in finite-dimensional spaces; the inequalities (3.18)–(3.20) are not used in this case. The constants in the estimates for the equivalent norms cannot be computed analytically due to the complicated structure of the norms. The constant in (3.28) depends on the dimension of* $\tilde{Y}_h$ *in addition to the constants in the estimates for the equivalent norms and those mentioned in Remark 3.25.*

3.6 Approximation result for the eigenvalues

We proved that the relations

$$|\lambda_0 - \lambda_h| + \|\underline{u}_0 - \underline{u}_h\|_{1,\Omega_S} \lesssim \Big\{\sum_{T\in\mathcal{T}_h} \eta_T^2\Big\}^{1/2} + \Big\{\sum_{T\in\mathcal{T}_h} \varepsilon_T^2\Big\}^{1/2}, \tag{3.29}$$

$$\Big\{\sum_{T\in\mathcal{T}_h} \eta_T^2\Big\}^{1/2} \lesssim |\lambda_0 - \lambda_h| + \|\underline{u}_0 - \underline{u}_h\|_{1,\Omega_S} + \Big\{\sum_{T\in\mathcal{T}_h} \varepsilon_T^2\Big\}^{1/2} \tag{3.30}$$

hold, where $\lambda_0 \in \mathbb{K}$ is a simple eigenvalue of problem (3.2), $\underline{u}_0 \in V$ is the corresponding eigenfunction and $[\lambda_h, \underline{u}_h] \in \mathbb{K} \times V_h$ is an approximation of $[\lambda_0, \underline{u}_0]$, see Theorems 3.17 and 3.23. The terms η_T and ε_T were introduced in (3.15) and (3.16) and denote the residual errors and the approximation errors, respectively.

The eigenvalues and eigenfunctions are estimated at the same time. It is known that the eigenfunctions dominate the convergence behavior while the eigenvalues converge faster, in fact. This means that the estimate (3.29) is suboptimal for the eigenvalues. This section is aimed at deriving from (3.29) an optimal a posteriori error estimate for the error $|\lambda_0 - \lambda_h|$. Similar approximation results for a lower bound are not known.

Recall that the eigenvalue problem (3.2) is equivalent to the eigenvalue problem for an operator pencil $\mathcal{B}(\cdot) : \mathbb{K} \to \mathcal{L}(V, V)$, see Section 2.3. We search for $\lambda \in \mathbb{K}$, $\underline{u} \in V$ such that

$$\mathcal{B}(\lambda)\underline{u} = 0.$$

The adjoint eigenvalue problem is given by: Find $\lambda \in \mathbb{K}$, $\underline{u}^\star \in V$ such that

$$\mathcal{B}^\star(\lambda)\underline{u}^\star = 0.$$

To compute an optimal upper bound for $|\lambda_0 - \lambda_h|$, we follow the ideas of Karma [49]. We learn from [49, Theorem 3] that $|\lambda_0 - \lambda_h| \lesssim (d_h d_h^\star)^{1/\kappa}$, where κ is the length of the longest Jordan chain at λ_0 and where d_h is the maximum distance of the generalized eigenfunctions from the space which is used for the discretization of the eigenvalue problem; $d_h^\star$ is defined similarly to d_h for the adjoint problem.

Fitted to our context, we have $\kappa = 1$ for a simple eigenvalue λ_0 and

$$d_h = d_h(\underline{u}) = \operatorname{dist}(\underline{u}, V_h) = \inf_{w \in V_h} \lbrack\!\lbrack \underline{u} - w \rbrack\!\rbrack_{1,\Omega_S}$$

provided that $\underline{u}$ is an eigenfunction corresponding to λ_0 with $\lbrack\!\lbrack \underline{u} \rbrack\!\rbrack_{1,\Omega_S} = 1$. In the previous calculations, we considered eigenfunctions $\underline{u}_0$ corresponding to λ_0 with $(\underline{u}_0, \underline{\bar{u}}_0)_H = 1$. Thus, $\underline{u} = \beta \underline{u}_0$ for some $\beta \in \mathbb{K}$ with $\lbrack\!\lbrack \beta \underline{u}_0 \rbrack\!\rbrack_{1,\Omega_S} = |\beta| \lbrack\!\lbrack \underline{u}_0 \rbrack\!\rbrack_{1,\Omega_S} = 1$. Then,

$$d_h = \inf_{w \in V_h} \lbrack\!\lbrack \beta \underline{u}_0 - w \rbrack\!\rbrack_{1,\Omega_S} \leq |\beta| \lbrack\!\lbrack \underline{u}_0 - \underline{u}_h \rbrack\!\rbrack_{1,\Omega_S} = 1/\lbrack\!\lbrack \underline{u}_0 \rbrack\!\rbrack_{1,\Omega_S} \lbrack\!\lbrack \underline{u}_0 - \underline{u}_h \rbrack\!\rbrack_{1,\Omega_S}.$$

Likewise, with $|\beta^\star| = 1/\lbrack\!\lbrack \underline{u}_0^\star \rbrack\!\rbrack_{1,\Omega_S} \in (0,1]$, the best approximation error for the adjoint problem is given by

$$d_h^\star = \inf_{w \in V_h} \lbrack\!\lbrack \beta^\star \underline{u}_0^\star - w \rbrack\!\rbrack_{1,\Omega_S} \leq \beta^\star \lbrack\!\lbrack \underline{u}_0^\star - \underline{u}_h \rbrack\!\rbrack_{1,\Omega_S} = 1/\lbrack\!\lbrack \underline{u}_0^\star \rbrack\!\rbrack_{1,\Omega_S} \lbrack\!\lbrack \underline{u}_0^\star - \underline{u}_h \rbrack\!\rbrack_{1,\Omega_S},$$

where $\underline{u}_0^\star$ is the eigenfunction of $\mathcal{B}^\star(\cdot)$ corresponding to λ_0.

Consequently, [49, Theorem 3] implies that

$$|\lambda_0 - \lambda_h| \lesssim \frac{1}{\lbrack\!\lbrack \underline{u}_0 \rbrack\!\rbrack_{1,\Omega_S}} \lbrack\!\lbrack \underline{u}_0 - \underline{u}_h \rbrack\!\rbrack_{1,\Omega_S} \cdot \frac{1}{\lbrack\!\lbrack \underline{u}_0^\star \rbrack\!\rbrack_{1,\Omega_S}} \lbrack\!\lbrack \underline{u}_0^\star - \underline{u}_h \rbrack\!\rbrack_{1,\Omega_S}. \tag{3.31}$$

In accordance with (3.15) and (3.16), we introduce the a posteriori error estimator $\eta_T^\star$ and the approximation error $\varepsilon_T^\star$ for the adjoint problem by

$$\begin{aligned} \eta_T^\star := \Big\{ h_{+,T}^2 \lbrack\!\lbrack \pi_{0,T} \underline{R}_T^\star \rbrack\!\rbrack_{0,T}^2 &+ \frac{1}{2} \sum_{E \in \mathcal{E}(T) \cap \mathcal{E}_{h,\Omega_S}} \frac{\lceil E \rceil}{\lceil T \rceil} h_{+,T}^2 \lbrack\!\lbrack \pi_{0,E} \underline{R}_E^\star \rbrack\!\rbrack_{0,E}^2 \\ &+ \sum_{E \in \mathcal{E}(T) \cap \mathcal{E}_{h,N}} \frac{\lceil E \rceil}{\lceil T \rceil} h_{+,T}^2 \lbrack\!\lbrack \pi_{0,E} \underline{R}_N^\star \rbrack\!\rbrack_{0,E}^2 \Big\}^{1/2}, \end{aligned} \tag{3.32}$$

$$\begin{aligned} \varepsilon_T^\star := \Big\{ h_{+,T}^2 \lbrack\!\lbrack \underline{R}_T^\star - \pi_{0,T} \underline{R}_T^\star \rbrack\!\rbrack_{0,T}^2 &+ \frac{1}{2} \sum_{E \in \mathcal{E}(T) \cap \mathcal{E}_{h,\Omega_S}} \frac{\lceil E \rceil}{\lceil T \rceil} h_{+,T}^2 \lbrack\!\lbrack \underline{R}_E^\star - \pi_{0,E} \underline{R}_E^\star \rbrack\!\rbrack_{0,E}^2 \\ &+ \sum_{E \in \mathcal{E}(T) \cap \mathcal{E}_{h,N}} \frac{\lceil E \rceil}{\lceil T \rceil} h_{+,T}^2 \lbrack\!\lbrack \underline{R}_N^\star - \pi_{0,E} \underline{R}_N^\star \rbrack\!\rbrack_{0,E}^2 \Big\}^{1/2}, \end{aligned} \tag{3.33}$$

where $\underline{R}_T^\star$, $\underline{R}_E^\star$ and $\underline{R}_N^\star$ are the element and edge residuals for the adjoint problem. With these definitions, we obtain an estimate similar to (3.29),

$$|\lambda_0 - \lambda_h| + \|\underline{u}_0^\star - \underline{u}_h^\star\|_{1,\Omega_S} \lesssim \Big\{\sum_{T\in\mathcal{T}_h} \eta_T^{\star 2}\Big\}^{1/2} + \Big\{\sum_{T\in\mathcal{T}_h} \varepsilon_T^{\star 2}\Big\}^{1/2}, \tag{3.34}$$

where $\underline{u}_0^\star$ satisfies $\mathcal{B}^\star(\lambda_0)\underline{u}_0^\star = 0$.

From (3.29) and (3.34), we obtain error estimates for the eigenfunctions and conclude an upper error bound for the approximation of the eigenvalues from (3.31). The results are summarized in the following theorem.

Theorem 3.27. *Let $\lambda_0 \in \mathbb{K}$ be a simple eigenvalue of problem (3.2). Let $\underline{u}_0 \in V$ be a corresponding eigenfunction with $(\underline{u}_0, \bar{\underline{u}}_0)_H = 1$ and let $\underline{u}_0^\star \in V$ be a corresponding eigenfunction of the adjoint eigenvalue problem satisfying $\mathcal{B}^\star(\lambda_0)\underline{u}_0^\star = 0$ and $(\underline{u}_0^\star, \bar{\underline{u}}_0^\star)_H = 1$. Furthermore, let $[\lambda_h, \underline{u}_h] \in \mathbb{K} \times V_h$ and $[\lambda_h, \underline{u}_h^\star] \in \mathbb{K} \times V_h$ with $(\underline{u}_h, \bar{\underline{u}}_h)_H = 1$ and $(\underline{u}_h^\star, \bar{\underline{u}}_h^\star)_H = 1$ be solutions to the discretized problems which are close enough to $[\lambda_0, \underline{u}_0]$ and $[\lambda_0, \underline{u}_0^\star]$ so that the assumptions of Lemma 2.4 are satisfied. Then, the following error estimates hold:*

$$\begin{aligned}
\|\underline{u}_0 - \underline{u}_h\|_{1,\Omega_S} &\lesssim \Big\{\sum_{T\in\mathcal{T}_h} \eta_T^2\Big\}^{1/2} + \Big\{\sum_{T\in\mathcal{T}_h} \varepsilon_T^2\Big\}^{1/2}, \\
\|\underline{u}_0^\star - \underline{u}_h^\star\|_{1,\Omega_S} &\lesssim \Big\{\sum_{T\in\mathcal{T}_h} \eta_T^{\star 2}\Big\}^{1/2} + \Big\{\sum_{T\in\mathcal{T}_h} \varepsilon_T^{\star 2}\Big\}^{1/2}, \\
|\lambda_0 - \lambda_h| &\lesssim \Big(\Big\{\sum_{T\in\mathcal{T}_h} \eta_T^2 + \varepsilon_T^2\Big\}^{1/2} \Big\{\sum_{T\in\mathcal{T}_h} \eta_T^{\star 2} + \varepsilon_T^{\star 2}\Big\}^{1/2} \\
&\lesssim \Big(\Big\{\sum_{T\in\mathcal{T}_h} \eta_T^2\Big\}^{1/2} + \Big\{\sum_{T\in\mathcal{T}_h} \varepsilon_T^2\Big\}^{1/2}\Big)\Big(\Big\{\sum_{T\in\mathcal{T}_h} \eta_T^{\star 2}\Big\}^{1/2} + \Big\{\sum_{T\in\mathcal{T}_h} \varepsilon_T^{\star 2}\Big\}^{1/2}\Big),
\end{aligned}$$

where η_T, ε_T, $\eta_T^\star$ and $\varepsilon_T^\star$ are defined in (3.15), (3.16), (3.32) and (3.33), respectively.

We saw in Section 3.4 that it is not necessary to project the residuals into a finite dimensional space in order to obtain an upper error bound, see Corollary 3.16. Using the error estimators

$$\tilde{\eta}_T := \Big\{h_{+,T}^2 \|\underline{R}_T\|_{0,\tilde{\omega}_T}^2 + \frac{1}{2} \sum_{E\in\mathcal{E}(T)\cap\mathcal{E}_{h,\Omega_S}} \frac{\lceil E\rceil}{\lceil T\rceil} h_{+,T}^2 \|\underline{R}_E\|_{0,E}^2 + \sum_{E\in\mathcal{E}(T)\cap\mathcal{E}_{h,N}} \frac{\lceil E\rceil}{\lceil T\rceil} h_{+,T}^2 \|\underline{R}_N\|_{0,E}^2\Big\}^{1/2}$$

and

$$\tilde{\eta}_T^\star := \Big\{h_{+,T}^2 \|\underline{R}_T^\star\|_{0,\tilde{\omega}_T}^2 + \frac{1}{2} \sum_{E\in\mathcal{E}(T)\cap\mathcal{E}_{h,\Omega_S}} \frac{\lceil E\rceil}{\lceil T\rceil} h_{+,T}^2 \|\underline{R}_E^\star\|_{0,E}^2 + \sum_{E\in\mathcal{E}(T)\cap\mathcal{E}_{h,N}} \frac{\lceil E\rceil}{\lceil T\rceil} h_{+,T}^2 \|\underline{R}_N^\star\|_{0,E}^2\Big\}^{1/2},$$

we obtain the following bounds.

Theorem 3.28. *Let $\lambda_0, \lambda_h \in \mathbb{C}$, $\underline{u}_0, \underline{u}_0^\star \in V$ and $\underline{u}_h, \underline{u}_h^\star \in V_h$ be as in Theorem 3.27. The following error estimates hold:*

$$\|\underline{u}_0 - \underline{u}_h\|_{1,\Omega_S} \lesssim \Big\{\sum_{T\in\mathcal{T}_h} \tilde{\eta}_T^2\Big\}^{1/2},$$

$$\|\lceil \underline{u}_0^\star - \underline{u}_h^\star \rceil\|_{1,\Omega_S} \lesssim \Big\{ \sum_{T\in\mathcal{T}_h} \tilde{\eta}_T^{\star 2} \Big\}^{1/2},$$

$$|\lambda_0 - \lambda_h| \lesssim \Big\{ \sum_{T\in\mathcal{T}_h} \tilde{\eta}_T^2 \Big\}^{1/2} \Big\{ \sum_{T\in\mathcal{T}_h} \tilde{\eta}_T^{\star 2} \Big\}^{1/2}.$$

Last but not least, it shall be noticed that the spherical nature of the problem influences the derivation of the error estimates only in the proofs of Lemmas 3.12, 3.13 and 3.14. The standard forms of these estimates for two-dimensional (Cartesian) domains are well-known results. This means that the theory of the previous sections is not restricted to spherical domains. The same error estimates hold for all eigenvalue problems of the type (2.13) on arbitrary two-dimensional manifolds provided that the assumptions on the operator pencil $\mathcal{B}(\cdot)$ (see Section 2.3) and on the triangulation of Ω_S (in particular, isotropy and comparable size of adjacent elements) are satisfied and that the assertions of Lemmas 3.12, 3.13 and 3.14 hold.

3.7 Constants

The constants in the error estimates are always of great interest. We traced the constants in the estimates of Sections 3.4, 3.5 and 3.6. To present all details of their derivation is a separate work and would go beyond the scope of this theses. This is why the result of this research will be given here without proof. For planar, two-dimensional domains, a strategy for the computation of the constants in the interpolation error estimates and in residual a posteriori error estimates was presented by Carstensen and Funken [22]. Their way of finding these constants is quite different from the strategies used in the present work and is restricted to real-valued functions. Moreover, most constants in [22] can be computed only for a limited number of special domains which consist of a union of half squares, whereas in this work more general domains are treated. The techniques used in the present work are mainly based on the consideration of star-shaped reference domains and the application of known estimates for planar domains. Admittedly, the estimates for the spherical domains are rather pessimistic (not at all optimal), because the transformation to reference domains and the weighted structure of the norms and operators on the sphere allow for very rough estimates for the constants only.

Most estimates are known only with respect to the (semi-)norms $\|\lceil\cdot\rceil\|_{0,\Omega_S}$ and $\lceil\cdot\rceil_{1,\Omega_S}$, while we consider a more general inner product $(\cdot,\cdot)_V$ with the associated norm $\|\underline{v}\|_V = \sqrt{(\underline{v},\underline{v})_V}$ on the space V of $\mathcal{H}^1$-functions. Thus, we need the constants $C_{1,V}$ and $C_{V,1}$ in the inequalities $\lceil\underline{v}\rceil_{1,\Omega_S} \le C_{1,V}\|\underline{v}\|_V$ and $\|\underline{v}\|_V \le C_{V,1}\|\lceil\underline{v}\rceil\|_{1,\Omega_S}$ in order to estimate the operator norms of F and $\mathrm{D}F(\lambda_0,\underline{u}_0)$.

Let $\mathcal{T}_h$ be a triangulation of Ω_S. Recall that the number of adjacent elements, which have one common vertex, is bounded from above by a constant $\mathcal{Z}$. If the mesh is constructed as suggested in Section 3.2, then $\mathcal{Z}$ is about 6 or 8. Let us assume for the moment that the constants $C_{\mathrm{I}_h,T}$ and $C_{\mathrm{I}_h,E}$ in the interpolation error estimates

$$\|\lceil\underline{v} - \mathrm{I}_h\,\underline{v}\rceil\|_{0,T} \le C_{\mathrm{I}_h,T} h_{+,T} \lceil\underline{v}\rceil_{1,\tilde{\omega}_T},$$

$$\|\lceil\underline{v} - \mathrm{I}_h\,\underline{v}\rceil\|_{0,E} \le C_{\mathrm{I}_h,E} \frac{\lceil E\rceil^{1/2}}{\lceil T\rceil^{1/2}} h_{+,T} \lceil\underline{v}\rceil_{1,\tilde{\omega}_E}.$$

are known, see Theorem 3.11. Then, the constant $C_{err,+}$ in

$$|\!|\!|[\lambda_0, \underline{u}_0] - [\lambda_h, \underline{u}_h]|\!|\!|_{\mathbb{K}\times V} \le C_{err,+} \Big\{ \sum_{T\in\mathcal{T}_h} \tilde{\eta}_T^2 \Big\}^{1/2}$$

(see Corollary 3.16) can be estimated by

$$C_{err,+} \le 2\|\mathrm{D}F([\lambda_0, \underline{u}_0])^{-1}\|_{\mathcal{L}(\mathbb{K}\times V^\star, \mathbb{K}\times V)} \max\{\max_{T\in\mathcal{T}_h} C_{\mathrm{I}_h,T}, \max_{E\in\mathcal{E}_h} C_{\mathrm{I}_h,E}\} \cdot (6\mathcal{Z} + 3)\, C_{1,V}.$$

The term $\|\mathrm{D}F([\lambda_0, \underline{u}_0])^{-1}\|_{\mathcal{L}(\mathbb{K}\times V^\star, \mathbb{K}\times V)}$ depends only on λ_0, $\underline{u}_0$, $\underline{u}_0^\star$ and on the definition of the operator pencil $\mathcal{B}$ (or of the function F), compare Lemma 2.12, and can therefore be treated as a constant. The constant in the estimate of Theorem 3.17 can be estimated by $C_{err,+}$ as well.

Closer considerations of the proof of Theorem 3.11 (see pages 68ff) show that

$$C_{\mathrm{I}_h,T} \le \sum_{\mathrm{x}\in\mathcal{N}(T)} C_P(\omega_\mathrm{x}) + \sum_{\mathrm{x}\in\mathcal{N}(T)\cap\mathcal{N}_{h,D}} C_{tr}(C_P(\omega_\mathrm{x}) + 1)\sqrt{C_{T,adj}} \cdot C_{h_+}$$

and

$$C_{\mathrm{I}_h,E} \le 2C_{tr} \sum_{\mathrm{x}\in\mathcal{N}(E)} (C_P(\omega_\mathrm{x}) + 1)\sqrt{C_{T,adj}} \cdot C_{h_+},$$

where $C_P(\omega_\mathrm{x})$ and C_{tr} are the constants in the Poincaré-type inequality (Lemma 3.13) and the trace theorem (Lemma 3.12), respectively, and where $C_{T,adj}$ and C_{h_+} are the constants in the relations $\lceil T_1 \rceil \le C_{T,adj} \lceil T_2 \rceil$ and $h_{+,T_1} \le C_{h_+} h_{+,T_2}$ for all pairs T_1, T_2 of adjacent elements. The constants $C_{T,adj}$ and C_{h_+} should have moderate values, when the assumption is satisfied that the size of adjacent elements does not change rapidly. When the mesh is generated as suggested in Algorithm 3.5, then both constants are about 1. In the course of an adaptive refinement, they may become larger. The constant C_{tr} in the trace theorem depends on the constant $\hat{c}_{tr}$ in the standard trace theorem on a reference element $\hat{T}$, for instance, $\hat{T} := \{(\hat{x}, \hat{y}) \mid 0 \le \hat{x} \le 1,\ 0 \le \hat{y} \le 1 - \hat{x}\}$, and on the constants in the norm inequalities (3.35)–(3.38) on page 62.

Moreover, the constant in

$$|\lambda_0 - \lambda_h| \lesssim \Big\{ \sum_{T\in\mathcal{T}_h} \tilde{\eta}_T^2 \Big\}^{1/2} \Big\{ \sum_{T\in\mathcal{T}_h} \tilde{\eta}_T^{\star\,2} \Big\}^{1/2},$$

see Theorem 3.28, can be estimated by $C_{err,+} C_{err,+}^\star C_{dd^\star} / (\|\underline{u}_0\|_{1,\Omega_S} \|\underline{u}_0^\star\|_{1,\Omega_S})$, where $C_{err,+}^\star$ is the constant in the upper error bound for the adjoint problem and $C_{dd^\star}$ is the constant in $|\lambda_0 - \lambda_h| \le C_{dd^\star} d_h d_h^\star$, compare Section 3.6. Because of $\tilde{\eta_T}^2 \le \eta_T^2 + \varepsilon_T^2$, the constant in the last estimate of Theorem 3.27 is the same.

The constant $C_{err,-}$ in the lower error bound

$$\Big\{ \sum_{T\in\mathcal{T}_h} \eta_T^2 \Big\}^{1/2} \le C_{err,-} \Big(|\!|\!|[\lambda, \underline{u}] - [\lambda_h, \underline{u}_h]|\!|\!|_{\mathbb{K}\times V} + \Big\{ \sum_{T\in\mathcal{T}_h} \varepsilon_T^2 \Big\}^{1/2} \Big)$$

is influenced by more factors. Let $C_{\eta_T,2}$ be the constant in the inequality (3.22). One will find that $C_{\eta_T,2} \le C_{iso} \max\{C_{R_T}, \frac{1}{2} C_{R_E}\}$ with C_{R_T} and C_{R_E} from (3.25) and (3.26), respectively, and C_{iso} from $h_{+,T} \le C_{iso} h_{-,T}$, see Remark 3.9. Then,

$$C_{err,-} \le \max\{12(\mathcal{Z} + 1) C_{V,1} C_{\eta_T,2} \|\mathrm{D}F([\lambda_0, \underline{u}_0])\|_{\mathcal{L}(\mathbb{K}\times V, \mathbb{K}\times V^\star)}, C_{1,V}(C_F + 3C_{tr}(C_F + 1))\}.$$

The constants C_{R_T} and C_{R_E} depend essentially on the constants in the estimates (3.18)–(3.20) for the bubble functions and on $C_{T,adj}$ and C_{h_+}. We omit the details of their derivation, because this would fill several more pages. One can show that

$$C_{R_T} \le (\frac{1}{4}C_{\mathfrak{b}_T,2} + C_{\mathfrak{b}_T,3})^{1/2}$$

and

$$C_{R_E} \le \Big((\frac{1}{4}C_{\mathfrak{b}_E,2} + C_{\mathfrak{b}_E,3})^{1/2} + \sqrt{C_{\mathfrak{b}_{T,E},2}}(\frac{1}{4}C_{\mathfrak{b}_T,2} + C_{\mathfrak{b}_T,3})^{1/2}\Big)(1 + C_{T,adj}C_{h_+}),$$

where $C_{\mathfrak{b}_T,2}$, $C_{\mathfrak{b}_T,3}$, $C_{\mathfrak{b}_E,2}$, $C_{\mathfrak{b}_E,3}$ and $C_{\mathfrak{b}_{T,E},2}$ are the constants in the estimates

$$\begin{array}{lllllll}
\int_T \mathfrak{b}_T^2 \,\mathrm{d}S & \le & C_{\mathfrak{b}_T,2}\lceil T\rceil, & & \|\lceil\nabla_S \mathfrak{b}_T\rceil\|_{0,T}^2 & \le & C_{\mathfrak{b}_T,3} h_{-,T}^{-2}\lceil T\rceil, \\
\int_E \mathfrak{b}_E^2 \,\mathrm{d}\sigma & \le & C_{\mathfrak{b}_E,2}\lceil E\rceil, & & \|\lceil\nabla_S \mathfrak{b}_E\rceil\|_{0,T}^2 & \le & C_{\mathfrak{b}_E,3} h_{-,T}^{-2}\lceil T\rceil \quad \forall T\subset\omega_E, \\
\int_T \mathfrak{b}_E^2 \,\mathrm{d}S & \le & C_{\mathfrak{b}_{T,E},2}\lceil T\rceil \quad \forall T\subset\omega_E. & & & &
\end{array}$$

We would like to remind of the fact that the constants in the estimates (3.18)–(3.20) are not obtained as easily as for planar domains due to special norms and nodal basis functions. Nevertheless, they can be computed approximately using the relations between the trigonometric functions and the results of Section 3.8.1, if the bubble functions are defined via the nodal basis suggested in Remark 3.10.

If Ω_S possesses a Dirichlet boundary part with positive measure, then the norms $\|\lceil \underline{w}_T\rceil\|_{1,T}$ and $\|\lceil \underline{w}_E\rceil\|_{1,\omega_E}$ in Lemma 3.20 (see page 51) can be replaced by the appropriate seminorms (since $\underline{w}_T$ and $\underline{w}_E$ vanish on ∂T and $\partial\omega_E$, respectively). If, in addition, the norm $\|\underline{u}\|_V = \lceil \underline{u}\rceil_{1,\Omega_S}$ is chosen, then $C_{V,1}$ and $C_{1,V}$ can be replaced by the constant 1 and the relations

$$C_{R_T} \le \sqrt{C_{\mathfrak{b}_T,3}}$$

and

$$C_{R_E} \le \Big(\sqrt{C_{\mathfrak{b}_T,3}} + \sqrt{C_{\mathfrak{b}_{T,E},2}}\sqrt{C_{\mathfrak{b}_T,3}}\Big)(1 + C_{T,adj}\cdot C_{h_+})$$

hold.

Like $\|\mathrm{D}F([\lambda_0,\underline{u}_0])^{-1}\|_{\mathcal{L}(\mathbb{K}\times V^\star,\mathbb{K}\times V)}$, the term $\|\mathrm{D}F([\lambda_0,\underline{u}_0])\|_{\mathcal{L}(\mathbb{K}\times V,\mathbb{K}\times V^\star)}$ can be treated as a constant. Estimates for these two constants are given in Lemma 2.12. Concerning the linear elasticity problem, it is particularly interesting how the constants depend on the material parameters. We present some dependencies in Section 4.3.2.

3.8 Proofs

This section contains the proofs of the interpolation error estimates (Theorem 3.11) and the auxiliary results: the trace theorem (Lemma 3.12) and the Poincaré-type inequality (Lemma 3.13) on spherical domains. These results are well-known for planar domains, see, for instance, [20]. The constants in the estimates depend on the geometry of the underlying domain. When we choose a fixed reference domain in $\mathbb{R}^2$, the constants in the appropriate estimates for this domain are independent of any discretization parameter. Thus, the aim is to transform our (spherical) domains and the corresponding norms and functions in the estimates to a planar reference domain and to apply the known results there. The backwards transformation yields the desired results. We will describe in the following section, how such a transformation can be performed.

3.8.1 Reference domains and transformation of norms

Let $\hat{T}$ be any fixed triangle in $\mathbb{R}^2$ with one marked edge $\hat{E}$. For each triangle $\tilde{T} \subset \mathbb{R}^2$ with a specified edge $\tilde{E}$, there is an affine linear map $F_{\tilde{T}} : \hat{T} \to \tilde{T}$ which maps $\hat{T}$ to $\tilde{T}$ and $\hat{E}$ to $\tilde{E}$. The triangle $\hat{T}$ is called *reference element* and the edge $\hat{E}$ is called *reference edge.* It is common to choose $\hat{T} := \{(\hat{x}, \hat{y}) \mid 0 \le \hat{x} \le 1,\ 0 \le \hat{y} \le 1 - \hat{x}\}$ and $\hat{E} := \{(\hat{x}, \hat{y}) \mid 0 \le \hat{x} \le 1,\ \hat{y} \equiv 0\}$. In the proof of Lemma 3.13, it is more convenient to use other reference elements.

Let $T \in \mathcal{T}_h$ be an element of the triangulation of the spherical domain $\Omega_\mathcal{S}$, and let $\Omega_\mathcal{S}$ be parametrized with spherical coordinates (φ, θ), see Sections 3.1 and 3.2. Since we consider open triangles T, each point $\underline{\mathbf{x}} = \underline{\mathbf{x}}(\varphi, \theta) \in T \subset \mathcal{S}^2$ can be uniquely identified with its coordinates $(\varphi, \theta) \in \tilde{T} \subset \tilde{\Omega}_\mathcal{S}$. For any function $v = v(\underline{\mathbf{x}})$ which is defined over T, let $\tilde{v} = \tilde{v}(\varphi, \theta) = v(\underline{\mathbf{x}}(\varphi, \theta))$ be the corresponding function which is defined over the element $\tilde{T}$ in the parameter domain. The integrals of these functions are related by

$$\int_T v(\underline{\mathbf{x}}(\varphi, \theta))\, \mathrm{d}\mathcal{S} = \int_{\tilde{T}} \tilde{v}(\varphi, \theta)\, \sin\theta\, \mathrm{d}\varphi\, \mathrm{d}\theta.$$

Furthermore, let $E \in \mathcal{E}(T)$ be an edge of T and let the corresponding edge $\tilde{E}$ of $\tilde{T}$ be given in the parametrized form $(\varphi(t), \theta(t))$ with $t \in (0, 1)$. We assumed that $\tilde{T}$ has straight-lined boundary so that the derivatives $\dot{\varphi}$ and $\dot{\theta}$ in the line element $\mathrm{d}\sigma$ are constant. The integrals over the edges are related by

$$\int_E v(\underline{\mathbf{x}}(\varphi, \theta))\, \mathrm{d}\sigma \quad = \quad \int_0^1 \tilde{v}(\varphi(t), \theta(t))\, \sqrt{\dot{\varphi}(t)^2 \sin^2\theta(t) + \dot{\theta}(t)^2}\, \mathrm{d}t.$$

In the following, we will consider all functions as functions of the variables φ and θ. For the sake of compactness of the notation, we will omit the $\sim$-symbol. It will always be clear from the context, whether the function under consideration is defined on the sphere or on the parameter domain.

The following lemma states the relations between the norms $\lceil\!|\cdot|\!\rceil$ on the sphere and the standard norms $\|\cdot\|$ on the two-dimensional reference element.

Lemma 3.29. *Let $T \in \mathcal{T}_h$. Each function $v \in \mathcal{H}^0(T)$ or $v \in \mathcal{H}^1(T)$ satisfies*

$$\lceil\!| v |\!\rceil_{0,T}^2 \sim \frac{\lceil T \rceil}{|\hat{T}|} \|\hat{v}\|_{0,\hat{T}}^2, \tag{3.35}$$

$$\lceil v \rceil_{1,T}^2 \lesssim \frac{\lceil T \rceil}{|\hat{T}|} h_{-,T}^{-2} |\hat{v}|_{1,\hat{T}}^2, \tag{3.36}$$

$$|\hat{v}|_{1,\hat{T}}^2 \lesssim \frac{|\hat{T}|}{\lceil T \rceil} h_{+,T}^2 \lceil v \rceil_{1,T}^2, \tag{3.37}$$

$$\lceil\!| v |\!\rceil_{0,E}^2 \sim \frac{\lceil E \rceil}{|\hat{E}|} \|\hat{v}\|_{0,\hat{E}}^2, \tag{3.38}$$

where $\hat{v}$ is the corresponding function on the reference domain.

An essential ingredient in the proof of Lemma 3.29 is relation (3.10) (see page 43). We will outline the proof of (3.35) and (3.38) and concentrate on the verification of the estimates (3.36) and (3.37), because this part is rather tricky. Beforehand, the attention shall be drawn again to Remark 3.3 (see page 39). Recall

$$\vartheta_{-,T} = \inf_{(\varphi,\theta)\in T} \sin\theta, \qquad \vartheta_{+,T} = \sup_{(\varphi,\theta)\in T} \sin\theta.$$

Note that, except for the poles and the line $\varphi \equiv 0$, all nodes $\mathrm{x} \in \mathcal{N}_h$ have a unique pendant in the parameter domain. By assumption, the elements $T \in \mathcal{T}_h$ do not cross the line $\varphi \equiv 0$ so that the vertices $\mathrm{x}_i \in \mathcal{N}(T)$ of each element T with $\vartheta_{-,T} > 0$ can uniquely be identified with coordinates (φ_i, θ_i), $i = 1, 2, 3$, where $\varphi = 2\pi$ is admissible as well. For pole elements ($\vartheta_{-,T} = 0$), only the node at the pole has no unique φ-coordinate. Pole elements appear as rectangles in the parameter domain, see page 42.

Proof of Lemma 3.29. Let us suppose first that T is not a pole element, that is, $\vartheta_{-,T} > 0$. There is an invertible map between $\tilde{T} \subset \tilde{\Omega}_S$ and $T \subset \Omega_S$, which identifies the vertices $\mathrm{x}_i \in \mathcal{N}(T)$ with the appropriate coordinates $(\varphi_i, \theta_i) \in [0, 2\pi] \times [0, \pi]$, $i = 1, 2, 3$, and which maps the (straight-lined) edges of $\tilde{T}$ to the (spherical) boundary of T.

The map $F_{\tilde{T}} : \hat{T} \to \tilde{T}$ from the reference element $\hat{T}$ to $\tilde{T}$ can be written in the form $(\varphi, \theta)^\top = F((\hat{x}, \hat{y})^\top) = \mathrm{B}(\hat{x}, \hat{y})^\top + \mathrm{b}$ with an appropriate transformation matrix B and a displacement vector b. If $\hat{T}$ is the standard reference element with the nodes $(0, 0)$, $(0, 1)$ and $(1, 0)$, then $F((\hat{x}, \hat{y})^\top) = \mathrm{B}(\hat{x}, \hat{y})^\top + \mathrm{b}$ with

$$\mathrm{B} = \begin{pmatrix} \varphi_2 - \varphi_1 & \varphi_3 - \varphi_1 \\ \theta_2 - \theta_1 & \theta_3 - \theta_1 \end{pmatrix}, \quad \mathrm{b} = \begin{pmatrix} \varphi_1 \\ \theta_1 \end{pmatrix}$$

maps $\hat{T}$ to $\tilde{T}$ and the reference edge $\hat{E} := \{(\hat{x}, \hat{y}) \mid 0 \le \hat{x} \le 1,\ \hat{y} \equiv 0\}$ to the edge $E \in \mathcal{E}(T)$ with $\mathcal{N}(E) = \{\mathrm{x}_1, \mathrm{x}_2\}$. In the proof of Lemma 3.13, we will use reference elements which differ from the standard reference element. For this reason, we consider a general transformation matrix

$$\mathrm{B} = \begin{pmatrix} \mathrm{B}_{11} & \mathrm{B}_{12} \\ \mathrm{B}_{21} & \mathrm{B}_{22} \end{pmatrix}$$

which maps all points in $\hat{T}$ to some point in the triangle $\tilde{T}$ up to a constant displacement.

The Jacobian of the transformation $\hat{T} \to \tilde{T}$ reads $|J| = |\det \mathrm{B}|$ so that

$$\int_{\tilde{T}} u(\varphi, \theta)\, \mathrm{d}\varphi\, \mathrm{d}\theta = |\det \mathrm{B}| \int_{\hat{T}} \hat{u}(\hat{x}, \hat{y})\, \mathrm{d}\hat{x}\, \mathrm{d}\hat{y}.$$

Note that $|\det \mathrm{B}| = |\tilde{T}|/|\hat{T}|$ and thus $\vartheta_{+,T}|\det \mathrm{B}| \sim \lceil T \rceil/|\hat{T}|$ due to relation (3.10). We obtain immediately that

$$\|\lceil v \rceil\|_{0,T}^2 = \int_{\tilde{T}} |v|^2 \sin\theta\, \mathrm{d}\varphi\, \mathrm{d}\theta \sim \vartheta_{+,T}|\det \mathrm{B}| \int_{\hat{T}} |\hat{v}|^2\, \mathrm{d}\hat{x}\, \mathrm{d}\hat{y} = \vartheta_{+,T}|\det \mathrm{B}| \|\hat{v}\|_{0,\hat{T}}^2 \sim \frac{\lceil T \rceil}{|\hat{T}|} \|\hat{v}\|_{0,\hat{T}}^2,$$

which proves (3.35). For the verification of the estimate (3.38) for the edge-norms, note that

$$\int_0^1 v(\varphi(t), \theta(t))\, \mathrm{d}t = \int_0^1 \hat{v}(\hat{x}(t), \hat{y}(t))\, \mathrm{d}t.$$

The length of $\hat{E}$ is given by $|\hat{E}| := (\dot{\hat{x}}^2 + \dot{\hat{y}}^2)^{1/2}$. The (reference) line element satisfies $\mathrm{d}\hat{\sigma} = |\hat{E}|\, \mathrm{d}t$. Using (3.10), we obtain that

$$\mathrm{d}\sigma = \sqrt{\dot{\varphi}^2 \sin^2\theta + \dot{\theta}^2}\, \mathrm{d}t \sim \sqrt{\dot{\varphi}^2 \vartheta_{+,T}^2 + \dot{\theta}^2}\, \mathrm{d}t \sim \lceil E \rceil\, \mathrm{d}t.$$

Consequently,

$$\frac{\|\lceil v\rceil\|^2_{0,E}}{\|\hat{v}\|^2_{0,\hat{E}}} = \frac{\displaystyle\int_0^1 v^2\,\mathrm{d}\sigma}{\displaystyle\int_0^1 \hat{v}^2\,\mathrm{d}\hat{\sigma}} \sim \frac{\lceil E\rceil \displaystyle\int_0^1 v^2\,\mathrm{d}t}{|\hat{E}|\displaystyle\int_0^1 \hat{v}^2\,\mathrm{d}t} = \frac{\lceil E\rceil}{|\hat{E}|},$$

which proves (3.38).

Now, we verify (3.36) and (3.37) for non-pole elements. We know that

$$\frac{\partial \hat{v}}{\partial \hat{x}} = \mathrm{B}_{11}\frac{\partial v}{\partial \varphi} + \mathrm{B}_{21}\frac{\partial v}{\partial \theta}, \qquad \frac{\partial \hat{v}}{\partial \hat{y}} = \mathrm{B}_{12}\frac{\partial v}{\partial \varphi} + \mathrm{B}_{22}\frac{\partial v}{\partial \theta}.$$

This means that

$$\hat{\nabla}\hat{v} = \begin{pmatrix} \frac{\partial \hat{v}}{\partial \hat{x}} \\ \frac{\partial \hat{v}}{\partial \hat{y}} \end{pmatrix} = \mathrm{B}^\top \cdot \begin{pmatrix} \frac{\partial v}{\partial \varphi} \\ \frac{\partial v}{\partial \theta} \end{pmatrix} = \mathrm{B}^\top \mathrm{D} \cdot \begin{pmatrix} \frac{1}{\sin\theta}\frac{\partial v}{\partial \varphi} \\ \frac{\partial v}{\partial \theta} \end{pmatrix}$$

with

$$\mathrm{D} = \begin{pmatrix} \sin\theta & 0 \\ 0 & 1 \end{pmatrix}.$$

Vice versa,

$$\begin{pmatrix} \frac{1}{\sin\theta}\frac{\partial v}{\partial \varphi} \\ \frac{\partial v}{\partial \theta} \end{pmatrix} = \mathrm{D}^{-1}\begin{pmatrix} \frac{\partial v}{\partial \varphi} \\ \frac{\partial v}{\partial \theta} \end{pmatrix} = \mathrm{D}^{-1}\mathrm{B}^{-\top} \cdot \begin{pmatrix} \frac{\partial \hat{v}}{\partial \hat{x}} \\ \frac{\partial \hat{v}}{\partial \hat{y}} \end{pmatrix} = \mathrm{D}^{-1}\mathrm{B}^{-\top} \cdot \hat{\nabla}\hat{v}.$$

Consequently,

$$|\hat{\nabla}\hat{v}|_2^2 \le \|\mathrm{B}^\top\mathrm{D}\|_2^2\,|\nabla_{\mathcal{S}} v|_2^2 \qquad \text{and} \qquad |\nabla_{\mathcal{S}} v|_2^2 \le \|\mathrm{D}^{-1}\mathrm{B}^{-\top}\|_2^2\,|\hat{\nabla}\hat{v}|_2^2,$$

where $\|\cdot\|_2$ is the spectral norm of the corresponding matrix and $|\cdot|_2$ is the Euclidean vector norm. We denote by $\delta_{\hat{T}}$ the diameter of the inscribed ball of $\hat{T}$. Due to the definition of the reference domain, $\delta_{\hat{T}}$ is a constant independent of the discretization parameter h. It holds that

$$\|\mathrm{B}^\top\mathrm{D}\|_2 = \|\mathrm{DB}\|_2 = \sup_{\hat{\mathrm{x}}\in\mathbb{R}^2\setminus\{0\}} \frac{|\mathrm{DB}\,\hat{\mathrm{x}}|_2}{|\hat{\mathrm{x}}|_2} = \sup_{\hat{\mathrm{x}}:|\hat{\mathrm{x}}|_2=\delta_{\hat{T}}} \frac{|\mathrm{DB}\,\hat{\mathrm{x}}|_2}{\delta_{\hat{T}}}.$$

Each line segment $\hat{\mathrm{x}} \subset \hat{T}$ with $|\hat{\mathrm{x}}|_2 = \delta_{\hat{T}}$ is mapped to some line segment x inside the element $\mathrm{B}\hat{T}$ and compressed by the factor $\sin\theta$ in the φ-direction after the application of D. Hence, $|\mathrm{DB}\,\hat{\mathrm{x}}|_2 = |\mathrm{x}|_2 \le \operatorname{diam}(\mathrm{DB}\,\hat{T}) \le \max\{h^{\mathcal{S}}_{\varphi,T}, h^{\mathcal{S}}_{\theta,T}\} = h_{+,T}$ (compare Remark 3.3). Thus, $|\hat{\nabla}\hat{v}|_2^2 \lesssim h_{+,T}^2|\nabla_{\mathcal{S}} v|_2^2$ and consequently,

$$|\hat{v}|^2_{1,\hat{T}} = \int_{\hat{T}} |\hat{\nabla}\hat{v}|_2^2\,\mathrm{d}\hat{x}\,\mathrm{d}\hat{y} \le \frac{1}{\vartheta_{-,T}|\det\mathrm{B}|}\int_{\tilde{T}} h_{+,T}^2|\nabla_{\mathcal{S}} v|_2^2 \sin\theta\,\mathrm{d}\varphi\,\mathrm{d}\theta \sim h_{+,T}^2\frac{|\hat{T}|}{\lceil T\rceil}\lceil v\rceil^2_{1,T},$$

which proves (3.37).

Now let δ_T be the diameter of the inscribed ball of $\mathrm{D}\tilde{T}$. Then,

$$\|\mathrm{D}^{-1}\mathrm{B}^{-\top}\|_2 = \|\mathrm{B}^{-1}\mathrm{D}^{-1}\|_2 = \sup_{\mathrm{x}\in\mathbb{R}^2\setminus\{0\}} \frac{|\mathrm{B}^{-1}\mathrm{D}^{-1}\,\mathrm{x}|_2}{|\mathrm{x}|_2} = \sup_{\mathrm{x}:|\mathrm{x}|_2=\delta_T} \frac{|\mathrm{B}^{-1}\mathrm{D}^{-1}\,\mathrm{x}|_2}{\delta_T}.$$

Each line segment $\mathrm{x} \in \mathrm{D}\tilde{T}$ with $|\mathrm{x}|_2 = \delta_T$ is mapped to some line segment $\hat{\mathrm{x}} = \mathrm{B}^{-1}\mathrm{D}^{-1}\mathrm{x}$ in $\hat{T}$. Hence, $|\mathrm{B}^{-1}\mathrm{D}^{-1}\mathrm{x}|_2 = |\hat{\mathrm{x}}|_2 \le \operatorname{diam}\hat{T} \sim 1$ and therefore, $|\nabla_{\mathcal{S}} v|_2^2 \lesssim \frac{1}{\delta_T^2}|\hat{\nabla}\hat{v}|_2^2$ and consequently,

$$\lceil v \rceil_{1,T}^2 \le \vartheta_{+,T}\int_{\tilde{T}} |\nabla_{\mathcal{S}} v|_2^2 \,\mathrm{d}\varphi\,\mathrm{d}\theta \lesssim \vartheta_{+,T}|\det \mathrm{B}| \int_{\hat{T}} \frac{1}{\delta_T^2}|\hat{\nabla}\hat{v}|_2^2 \,\mathrm{d}\hat{x}\,\mathrm{d}\hat{y} \sim \frac{1}{\delta_T^2}\frac{\lceil T \rceil}{|\hat{T}|}|\hat{v}|_{1,\hat{T}}^2.$$

Due to Remark 3.3, the dimensions of $\mathrm{D}\tilde{T}$ are given by $h_{\varphi,T}^{\mathcal{S}}$ and $h_{\theta,T}^{\mathcal{S}}$. Owing to the assumption (3.7) on the mesh (see page 42), we have that $1/\delta_T \lesssim h_{-,T}^{-1}$, which proves (3.36).

Now suppose that T is a pole element, this means that $\vartheta_{-,T} = 0$ and that the relation (3.10) cannot be applied. By assumption, one of the nodes x_i of T, $i = 1, 2, 3$, corresponds to the pole and the other two nodes have the same θ-coordinate. Without loss of generality, let x_3 be the pole node; this means that $\sin\theta_3 = 0$ and $\theta_2 = \theta_1$. Let the nodes x_1 and x_2 be enumerated so that the corresponding φ-coordinates satisfy $\varphi_1 \le \varphi_2$. The node x_3 has no unique pendant in the parameter domain; the corresponding φ-coordinate φ_3 varies between φ_1 and φ_2. Note that $h_{\varphi,T} = \varphi_2 - \varphi_1 > 0$, $h_{\theta,T} = \theta_1 = \theta_2 = h_{\theta,T}^{\mathcal{S}}$, $\vartheta_{+,T} = \sin h_{\theta,T}$.

As the reference domain, we choose a sector $\hat{T}$ of the unit circle with opening angle γ, for instance, $\gamma = \frac{\pi}{2}$. The element T is mapped to $\hat{T}$ by the transformation

$$\hat{x} = \vartheta_{+,T}^{-1}\cos\frac{\gamma(\varphi - \varphi_1)}{h_{\varphi,T}}\sin\theta, \qquad \hat{y} = \vartheta_{+,T}^{-1}\sin\frac{\gamma(\varphi - \varphi_1)}{h_{\varphi,T}}\sin\theta.$$

The Jacobian of this transformation reads $|J| = \gamma(h_{\varphi,T}\vartheta_{+,T}^2)^{-1}\sin\theta|\cos\theta|$. The partial derivatives are related by

$$\left[\left(\frac{\gamma^{-1}h_{\varphi,T}\vartheta_{+,T}}{\sin\theta}\frac{\partial v}{\partial\varphi}\right)^2 + \left(\frac{\vartheta_{+,T}}{\cos\theta}\frac{\partial v}{\partial\theta}\right)^2\right] = \left(\frac{\partial\hat{v}}{\partial\hat{x}}\right)^2 + \left(\frac{\partial\hat{v}}{\partial\hat{y}}\right)^2. \tag{3.39}$$

We required that $h_{\theta,T} \le \frac{\pi}{4}$ at the poles. Hence, $|\cos\theta| \sim 1$ and $\sin\theta \sim \theta$. Note that $|\hat{T}| = \gamma/2$ and $\lceil T \rceil = h_{\varphi,T}|\cos\theta_1 - \cos\theta_3| = 2h_{\varphi,T}\sin^2(|\theta_1 - \theta_3|/2) \sim h_{\varphi,T}\vartheta_{+,T}^2$ for $\theta_3 \in \{0, \pi\}$. Moreover, $\gamma \sim 1$, $h_{\varphi,T}\vartheta_{+,T} = h_{\varphi,T}^{\mathcal{S}}$ and $\vartheta_{+,T} \sim h_{\theta,T}^{\mathcal{S}}$. Consequently,

$$\begin{aligned}
\|\hat{v}\|_{0,\hat{T}}^2 &= \int_{\hat{T}} \hat{v}^2 \,\mathrm{d}\hat{x}\,\mathrm{d}\hat{y} = \int_T |J|\, v^2 \,\mathrm{d}\varphi\,\mathrm{d}\theta \sim \gamma(h_{\varphi,T}\vartheta_{+,T}^2)^{-1}\lceil\!\lceil v \rceil\!\rceil_{0,T}^2 \sim \frac{|\hat{T}|}{\lceil T \rceil}\lceil\!\lceil v \rceil\!\rceil_{0,T}^2,\\
|\hat{v}|_{1,\hat{T}}^2 &= \int_{\hat{T}} \left(\frac{\partial\hat{v}}{\partial\hat{x}}\right)^2 + \left(\frac{\partial\hat{v}}{\partial\hat{y}}\right)^2 \mathrm{d}\hat{x}\,\mathrm{d}\hat{y}\\
&= \int_T |J|\left[\left(\frac{\gamma^{-1}h_{\varphi,T}\vartheta_{+,T}}{\sin\theta}\frac{\partial v}{\partial\varphi}\right)^2 + \left(\frac{\vartheta_{+,T}}{\cos\theta}\frac{\partial v}{\partial\theta}\right)^2\right]\mathrm{d}\varphi\,\mathrm{d}\theta\\
&\sim \gamma(h_{\varphi,T}\vartheta_{+,T}^2)^{-1}\left[(h_{\varphi,T}^{\mathcal{S}})^2\lceil\!\lceil \frac{1}{\sin\theta}\frac{\partial v}{\partial\varphi}\rceil\!\rceil_{0,T}^2 + (h_{\theta,T}^{\mathcal{S}})^2\lceil\!\lceil \frac{\partial v}{\partial\theta}\rceil\!\rceil_{0,T}^2\right]\\
&\lesssim \max\{h_{\varphi,T}^2\vartheta_{+,T}^2, h_{\theta,T}^2\}|\hat{T}|/\lceil T \rceil\,\lceil v \rceil_{1,T}^2 = h_{+,T}^2|\hat{T}|/\lceil T \rceil\,\lceil v \rceil_{1,T}^2,\\
|\hat{v}|_{1,\hat{T}}^2 &\gtrsim \min\{h_{\varphi,T}^2\vartheta_{+,T}^2, h_{\theta,T}^2\}|\hat{T}|/\lceil T \rceil\,\lceil v \rceil_{1,T}^2 = h_{-,T}^2|\hat{T}|/\lceil T \rceil\,\lceil v \rceil_{1,T}^2,\\
\lceil v \rceil_{1,T}^2 &\lesssim h_{-,T}^{-2}\lceil T \rceil/|\hat{T}|\,|\hat{v}|_{1,\hat{T}}^2.
\end{aligned}$$

Furthermore, let $E \in \mathcal{E}(T)$ and let $\hat{E}$ be the corresponding reference edge which is determined uniquely by the given transformation. To verify that the relation $\lceil\!\lceil v \rceil\!\rceil_{0,E}^2 = \lceil E \rceil/|\hat{E}| \cdot \|\hat{v}\|_{0,\hat{E}}^2$ holds, we distinguish two cases.

If E is a vertical edge, then $\varphi \equiv const.$ and $\theta = \theta(t) = \theta_3 + t \cdot (\theta_1 - \theta_3)$. Hence,

$$\lceil E \rceil = \int_0^1 \mathrm{d}\sigma = \int_0^1 \sqrt{\dot{\varphi}^2 \sin^2\theta + \dot{\theta}^2}\,\mathrm{d}t = \sqrt{(\theta_1 - \theta_3)^2} = h_{\theta,T} = h^{\mathcal{S}}_{\theta,T}.$$

The corresponding reference edge has the length $|\hat{E}| = 1 = (\dot{\hat{x}}^2 + \dot{\hat{y}}^2)^{1/2}$ (the radius of the unit circle), and the assertion follows from $\|\lceil v \rceil\|^2_{0,E} / \lceil E \rceil = \int_0^1 v^2\,\mathrm{d}t = \int_0^1 \hat{v}^2\,\mathrm{d}t = \|\hat{v}\|_{0,\hat{E}} / |\hat{E}|$.

If E is a horizontal edge, then $\varphi = \varphi(t) = \varphi_1 + t(\varphi_2 - \varphi_1)$ and $\theta \equiv \theta_1 \equiv const.$ Hence,

$$\lceil E \rceil = \int_0^1 \mathrm{d}\sigma = \int_0^1 \sqrt{\dot{\varphi}^2 \sin^2\theta + \dot{\theta}^2}\,\mathrm{d}t = \int_0^1 h_{\varphi,T} \sin\theta_1\,\mathrm{d}t = h_{\varphi,T}\vartheta_{+,T} = h^{\mathcal{S}}_{\varphi,T}.$$

The corresponding reference edge is the curved line of the circular sector with the length $|\hat{E}| = 2\pi \cdot \gamma/(2\pi) = \gamma$. Again, the assertion follows from $\|\lceil v \rceil\|^2_{0,E} / \lceil E \rceil = \int_0^1 v^2\,\mathrm{d}t = \int_0^1 \hat{v}^2\,\mathrm{d}t = \|\hat{v}\|_{0,\hat{E}} / |\hat{E}|$. □

3.8.2 Proofs of the results of Section 3.3

In this subsection, we prove the interpolation error estimates of Theorem 3.11 and the auxiliary results of Section 3.3 which are involved in the verification of the upper error bound (Theorem 3.17). The definition of the norms allows us to consider each component of the vector functions as separate scalar functions so that the relations (3.35)–(3.38) can be employed to prove the desired estimates.

Proof of Lemma 3.12 (Trace theorem for functions on the sphere). Assertion: Let $T \in \mathcal{T}_h$ and $E \in \mathcal{E}_h(T)$. Each $\underline{v} \in [\mathcal{H}^1(T)]^d$ satisfies

$$\|\lceil \underline{v} \rceil\|_{0,E} \lesssim \frac{\lceil E \rceil^{1/2}}{\lceil T \rceil^{1/2}} \left(\|\lceil \underline{v} \rceil\|^2_{0,T} + h^2_{+,T} \lceil \underline{v} \rceil^2_{1,T} \right)^{1/2}.$$

If $d > 1$, let v_i, $i = 1, \ldots, d$, be the components of the vector function $\underline{v} = \sum_{i=1}^3 v_i \mathbf{e}_i$ in the Cartesian basis. Suppose that the trace theorem holds for scalar functions. Then,

$$\|\lceil \underline{v} \rceil\|^2_{0,E} = \sum_{i=1}^3 \|\lceil v_i \rceil\|^2_{0,E} \lesssim \sum_{i=1}^3 \frac{\lceil E \rceil}{\lceil T \rceil} \left(\|\lceil v_i \rceil\|^2 + h^2_{+,T} \lceil v_i \rceil^2_{1,T} \right) = \frac{\lceil E \rceil}{\lceil T \rceil} \left(\|\lceil \underline{v} \rceil\|^2 + h^2_{+,T} \lceil \underline{v} \rceil^2_{1,T} \right).$$

Thus, it is sufficient to prove the assertion for scalar functions.

The standard trace theorem has the form

$$\|\hat{v}\|_{0,\hat{E}} \leq \hat{c}_{tr} \|\hat{v}\|_{1,\hat{T}}, \tag{3.40}$$

where $\hat{T} \subset \mathbb{R}^2$, $\hat{E} \subset \partial\hat{T}$ and $\hat{c}_{tr}$ depends on the length $|\hat{E}|$ of $\hat{E}$ and the size $|\hat{T}|$ of $\hat{T}$ and is therefore independent of the triangulation and the discretization parameter h. We employ the relations (3.38), (3.35) and (3.37) to conclude that

$$\begin{aligned}
\|\lceil v \rceil\|^2_{0,E} &\sim \frac{\lceil E \rceil}{|\hat{E}|} \|\hat{v}\|^2_{0,\hat{E}} \leq \hat{c}^2_{tr} \frac{\lceil E \rceil}{|\hat{E}|} (\|\hat{v}\|^2_{0,\hat{T}} + |\hat{v}|^2_{1,\hat{T}}) \\
&\lesssim \frac{\lceil E \rceil}{|\hat{E}|} \left(\frac{|\hat{T}|}{\lceil T \rceil} \|\lceil v \rceil\|^2_{0,T} + h^2_{+,T} \frac{|\hat{T}|}{\lceil T \rceil} \lceil v \rceil^2_{1,T} \right) \\
&\sim \frac{\lceil E \rceil}{\lceil T \rceil} (\|\lceil v \rceil\|^2_{0,T} + h^2_{+,T} \lceil v \rceil^2_{1,T}).
\end{aligned}$$

□

Proof of Lemma 3.13. Assertion: Let $\mathrm{x} \in \mathcal{N}_h$ and $T \in \omega_\mathrm{x}$. The relation

$$\|\lceil \underline{v} - \pi_{0,\omega_\mathrm{x}}\underline{v}\rceil\|_{0,T} \le \|\lceil \underline{v} - \pi_{0,\omega_\mathrm{x}}\underline{v}\rceil\|_{0,\omega_\mathrm{x}} \lesssim h_{+,T}\lceil \underline{v}\rceil_{1,\omega_\mathrm{x}}$$

holds for all $\underline{v} \in [\mathcal{H}^1(\omega_\mathrm{x})]^d$.

If $d > 1$, let v_i, $i = 1, \ldots, d$, be the components of the vector function $\underline{v} = \sum_{i=1}^3 v_i \mathbf{e}_i$ in the Cartesian basis. Suppose that the assertion holds for scalar functions. Then,

$$\|\lceil \underline{v} - \pi_{0,T}\underline{v}\rceil\|_{0,\omega_\mathrm{x}}^2 = \sum_{i=1}^3 \|\lceil v_i - \pi_{0,T}v_i\rceil\|_{0,\omega_\mathrm{x}}^2 \lesssim \sum_{i=1}^3 h_{+,T}^2\lceil v_i\rceil_{1,\omega_\mathrm{x}}^2 = h_{+,T}^2\lceil \underline{v}\rceil_{1,\omega_\mathrm{x}}^2.$$

Thus, it is sufficient to prove the assertion for scalar functions.

It was proven by Dupont and Scott [35], see also Brenner and Scott [20], that for each domain $\hat{\omega} \subset \mathbb{R}^2$ which is star-shaped with respect to a ball B, there is a constant $\hat{c}_B = \hat{c}_B(\hat{\omega}) > 0$ so that

$$\forall \hat{v} \in H^1(\hat{\omega}) \quad \exists \hat{\tau} \in \mathcal{P}_{0|\hat{\omega}} : \qquad \|\hat{v} - \hat{\tau}\|_{0,\hat{\omega}} \le \hat{c}_B|\hat{v}|_{1,\hat{\omega}}. \tag{3.41}$$

The constant $\hat{c}_B(\hat{\omega})$ depends on the ratio of the diameter of $\hat{\omega}$ and the radius of the largest ball B with respect to which $\hat{\omega}$ is star-shaped. Using this result, we will prove that

$$\forall v \in \mathcal{H}^1(\omega_\mathrm{x}) \quad \exists \tau = \tau(v) \in \mathcal{P}_{0|\omega_\mathrm{x}} : \quad \|\lceil v - \tau\rceil\|_{0,\omega_\mathrm{x}} \lesssim h_{+,T}\lceil v\rceil_{1,\omega_\mathrm{x}}. \tag{3.42}$$

Since the operator $\pi_{0,\omega_\mathrm{x}}$ projects constant functions to themselves, in particular $\pi_{0,\omega_\mathrm{x}}\tau = \tau$, the assertion of Lemma 3.13 follows then from $\|\lceil \pi_{0,\omega_\mathrm{x}}(v-\tau)\rceil\|_{0,\omega_\mathrm{x}} \le \|\lceil v - \tau\rceil\|_{0,\omega_\mathrm{x}}$ and from

$$\begin{aligned}\|\lceil v - \pi_{0,\omega_\mathrm{x}}v\rceil\|_{0,\omega_\mathrm{x}} &= \|\lceil (v-\tau) - \pi_{0,\omega_\mathrm{x}}(v-\tau)\rceil\|_{0,\omega_\mathrm{x}} \le \|\lceil v-\tau\rceil\|_{0,\omega_\mathrm{x}} + \|\lceil \pi_{0,\omega_\mathrm{x}}(v-\tau)\rceil\|_{0,\omega_\mathrm{x}} \\ &\le 2\|\lceil v-\tau\rceil\|_{0,\omega_\mathrm{x}} \lesssim h_{+,T}\lceil v\rceil_{1,\omega_\mathrm{x}}.\end{aligned}$$

To show (3.42), we construct a star-shaped reference domain. Suppose first that $\mathrm{x} \in \mathcal{N}_h$ is a node so that ω_x does not touch one of the poles (that is, $\vartheta_{-,\omega_\mathrm{x}} > 0$) and therefore relation (3.10) on page 43 can be applied. Let N_x be the number of triangles in ω_x. By assumption, N_x is bounded independently of h. Following the idea of Kunert [60], we define $\hat{\omega}_\mathrm{x}$ as follows: If $\mathrm{x} \notin \partial\Omega_S$, then ω_x is transformed to a regular polygon with its vertices on the unit circle, otherwise to a polygon in the quarter unit circle, see Figure 3.5. In each case, $\hat{\omega}_\mathrm{x}$ can be divided into N_x equal triangles $\hat{T}_i$, $i = 1, \ldots, N_\mathrm{x}$, which correspond to the elements T_i of ω_x, respectively. Furthermore, each element T_i of ω_x corresponds to an element $\tilde{T}_i$ in the parameter domain, so that the transformation from $\hat{\omega}_\mathrm{x}$ to ω_x can be expressed by a *continuous*, piecewise affine linear map $F(\hat{\omega})$ which is composed of affine linear maps $F_i : \hat{T}_i \to \tilde{T}_i$. The map F is constructed so that each pair of elements $\tilde{T}_i, \tilde{T}_j$ with a common edge E is transformed to the pair $\hat{T}_i, \hat{T}_j$ with the common edge $\hat{E} = F^{-1}(E)$ so that common nodes are mapped to common nodes. Thus, each function $v \in \mathcal{H}^1(\omega)$ is the image of a function $\hat{v} \in H^1(\hat{\omega})$, $v(\omega) = v(F(\hat{\omega})) = \hat{v}(\hat{\omega})$, cf. [91].

The relations (3.35) and (3.37) are true for each $T_i \subset \omega_\mathrm{x}$. The size of adjacent elements does not change rapidly, so that $\lceil T_i\rceil \sim \lceil T\rceil$ and $h_{+,T_i} \sim h_{+,T}$ for each fixed element $T \subset \omega_\mathrm{x}$ and all $T_i \subset \omega_\mathrm{x}$. Moreover, $|\hat{T}_i| = |\hat{T}|$ for all $i = 1, \ldots, N_\mathrm{x}$. We conclude for any constant function τ over ω_x that

$$\begin{aligned}\|\lceil v - \tau\rceil\|_{0,\omega_\mathrm{x}}^2 &= \sum_{i=1}^{N_\mathrm{x}} \|\lceil v-\tau\rceil\|_{0,T_i}^2 \sim \sum_{i=1}^{N_\mathrm{x}} \frac{\lceil T_i\rceil}{|\hat{T}_i|}\|\hat{v} - \hat{\tau}\|_{0,\hat{T}_i}^2 \sim \frac{\lceil T\rceil}{|\hat{T}|}\|\hat{v} - \hat{\tau}\|_{0,\hat{\omega}_\mathrm{x}}^2, \\ |\hat{v}|_{1,\hat{\omega}_\mathrm{x}}^2 &= \sum_{i=1}^{N_\mathrm{x}} |\hat{v}|_{1,\hat{T}_i}^2 \lesssim \sum_{i=1}^{N_\mathrm{x}} \frac{|\hat{T}_i|}{\lceil T_i\rceil} h_{+,T_i}^2\lceil v\rceil_{1,T_i}^2 \sim \frac{|\hat{T}|}{\lceil T\rceil} h_{+,T}^2\lceil v\rceil_{1,\omega_\mathrm{x}}^2.\end{aligned}$$

By construction, the domain $\hat{\omega}_{\mathrm{x}}$ is star-shaped with respect to a ball, so that the estimate (3.41) holds for some constant $\hat{\tau}$. In combination with the previous estimates, we conclude that

$$\lceil v - \tau \rceil\!\rceil_{0,\omega_{\mathrm{x}}}^2 \lesssim \frac{\lceil T \rceil}{|\hat{T}|} |\hat{v}|_{1,\hat{\omega}_{\mathrm{x}}}^2 \lesssim h_{+,T}^2 \lceil v \rceil_{1,\omega_{\mathrm{x}}}^2,$$

which proves (3.42) for non-pole elements.

Now suppose that ω_{x} touches a pole (that is, $\vartheta_{-,\omega_{\mathrm{x}}} = 0$). By assumption, there is only a limited number of such domains. We assumed that ω_{x} touches a crack tip or a boundary corner at most directly at a pole (see page 42) and that Ω_S is a regular domain, so that ω_{x} is star-shaped with respect to a ball.

We use the same transformation as for single pole elements (see page 65), where we replace T by ω_{x}. This means, let $\varphi_1 = \inf_{\underline{\mathrm{x}}(\varphi,\theta)\in\omega_{\mathrm{x}}} \varphi$, $\varphi_2 = \sup_{\underline{\mathrm{x}}(\varphi,\theta)\in\omega_{\mathrm{x}}} \varphi$, $h_{\varphi,\omega_{\mathrm{x}}} = \varphi_2 - \varphi_1$, and

$$\hat{x} = \vartheta_{+,\omega_{\mathrm{x}}}^{-1} \cos \frac{\gamma(\varphi - \varphi_1)}{h_{\varphi,\omega_{\mathrm{x}}}} \sin\theta, \qquad \hat{y} = \vartheta_{+,\omega_{\mathrm{x}}}^{-1} \sin \frac{\gamma(\varphi - \varphi_1)}{h_{\varphi,\omega_{\mathrm{x}}}} \sin\theta.$$

Like for non-pole elements, ω_{x} consists of a limited number of elements T_i, $i = 1, \ldots, N_{\mathrm{x}}$, and $\hat{\omega}_{\mathrm{x}}$ is a part of the unit circle, in particular, of the circular sector with opening angle γ, with the corresponding elements $\hat{T}_i$. Figure 3.6 displays the possible structures of $\hat{\omega}_{\mathrm{x}}$ depending on the position of the node x. Similarly to the argumentation in the proof of Lemma 3.29, we obtain that

$$\begin{aligned} \lceil v \rceil\!\rceil_{0,\omega_{\mathrm{x}}}^2 &\sim \gamma^{-1}(h_{\varphi,\omega_{\mathrm{x}}} \vartheta_{+,\omega_{\mathrm{x}}}^2) \|\hat{v}\|_{0,\hat{\omega}_{\mathrm{x}}}^2, \\ |\hat{v}|_{1,\hat{\omega}_{\mathrm{x}}}^2 &\lesssim \gamma(h_{\varphi,\omega_{\mathrm{x}}} \vartheta_{+,\omega_{\mathrm{x}}}^2)^{-1} \max\{\gamma^{-2} h_{\varphi,\omega_{\mathrm{x}}}^2 \vartheta_{+,\omega_{\mathrm{x}}}^2, h_{\theta,\omega_{\mathrm{x}}}^2\} \lceil v \rceil_{1,\omega_{\mathrm{x}}}^2. \end{aligned}$$

Since $\hat{\omega}_{\mathrm{x}}$ is star-shaped, too, the relation (3.41) can be applied. Thus, there is a constant function τ so that

$$\begin{aligned} \lceil v - \tau \rceil_{0,\omega_{\mathrm{x}}}^2 &\sim \gamma^{-1}(h_{\varphi,\omega_{\mathrm{x}}} \vartheta_{+,\omega_{\mathrm{x}}}^2) \|\hat{v} - \hat{\tau}\|_{0,\hat{\omega}_{\mathrm{x}}}^2 \leq \hat{c}_B^2 \gamma^{-1}(h_{\varphi,\omega_{\mathrm{x}}} \vartheta_{+,\omega_{\mathrm{x}}}^2) |\hat{v}|_{1,\omega_{\mathrm{x}}}^2 \\ &\lesssim \max\{\gamma^{-2} h_{\varphi,\omega_{\mathrm{x}}}^2 \vartheta_{+,\omega_{\mathrm{x}}}^2, h_{\theta,\omega_{\mathrm{x}}}^2\} \lceil v \rceil_{1,\omega_{\mathrm{x}}}^2. \end{aligned}$$

Since $\gamma \sim 1$, $h_{\varphi,\omega_{\mathrm{x}}} \sim h_{\varphi,T}$, $\vartheta_{+,\omega_{\mathrm{x}}} \sim \vartheta_{+,T}$ and $h_{\theta,\omega_{\mathrm{x}}} \sim h_{\theta,T}$ for each $T \subset \omega_{\mathrm{x}}$, the estimate (3.42) is proven also for pole elements, which completes the proof of Lemma 3.13.

□

Remark 3.30. *Contrary to the conditions in [6], it is not necessary to assume that all elements have at least one edge which is parallel to the φ-axis. This condition is exploited only for the elements near the poles, that is, for $T \in \omega_{\mathrm{x}}$ with $\vartheta_{-,\omega_{\mathrm{x}}} = 0$.*

Remark 3.31. *The main idea of the proof of Lemma 3.13 is to find reference domains which are star-shaped with respect to a ball. We remark that the pendant of the domain ω_{x} in the parameter domain is not necessarily connected, let alone star-shaped, if the φ-coordinate of x equals 0 (or 2π). Since we use piecewise affine linear maps, however, this fact does not cause any problems; the resulting reference domain is star-shaped, nevertheless.*

Proof of Theorem 3.11 (Interpolation error estimates). Assertion: The interpolation operator satisfies

$$\begin{aligned} \lceil \underline{v} - \mathrm{I}_h \underline{v} \rceil\!\rceil_{0,T} &\lesssim h_{+,T} \lceil \underline{v} \rceil_{1,\tilde{\omega}_T} \quad \forall \underline{v} \in [\mathcal{H}^1(\tilde{\omega}_T)]^d \cap V, \\ \lceil \underline{v} - \mathrm{I}_h \underline{v} \rceil\!\rceil_{0,E} &\lesssim \frac{\lceil E \rceil^{1/2}}{\lceil T \rceil^{1/2}} h_{+,T} \lceil \underline{v} \rceil_{1,\tilde{\omega}_E} \quad \forall \underline{v} \in [\mathcal{H}^1(\tilde{\omega}_E)]^d \cap V. \end{aligned}$$

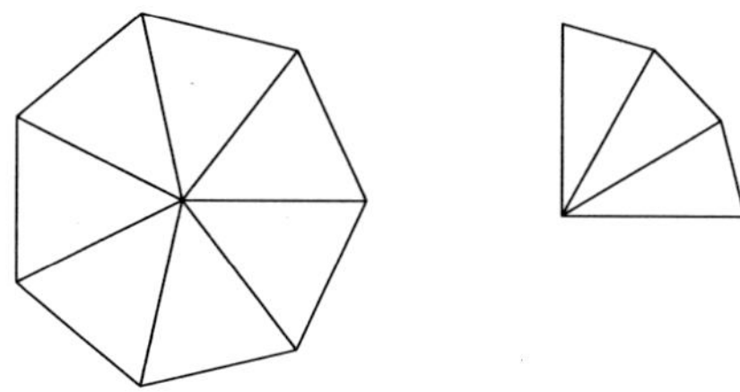

Figure 3.5: Reference patches for ω_{x} with $\vartheta_{-,\omega_{\mathrm{x}}} > 0$; left: $\mathrm{x} \notin \partial\Omega_{\mathcal{S}}$, right: $\mathrm{x} \in \partial\Omega_{\mathcal{S}}$

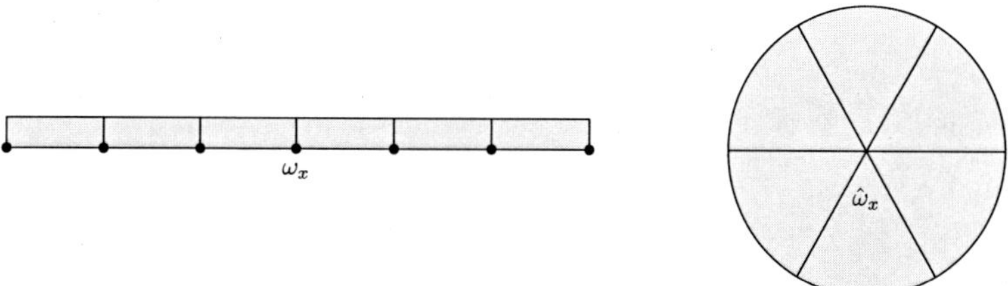

x is a pole and the entire neighborhood of x belongs to $\Omega_{\mathcal{S}}$ ($\rightarrow \gamma = 2\pi$);

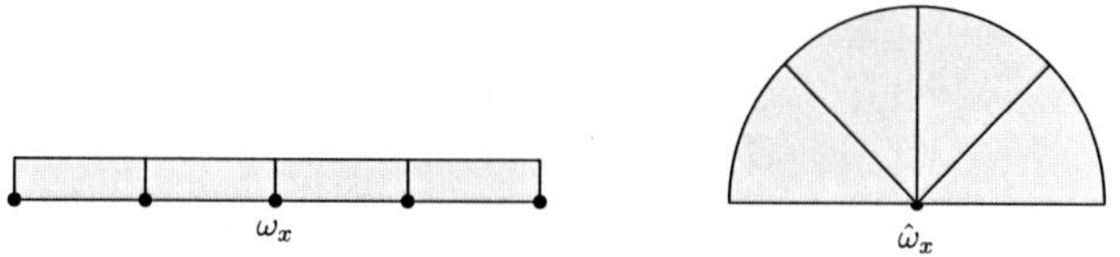

x is a pole and only a part of the neighborhood of x belongs to $\Omega_{\mathcal{S}}$ ($\rightarrow \gamma = \pi$);

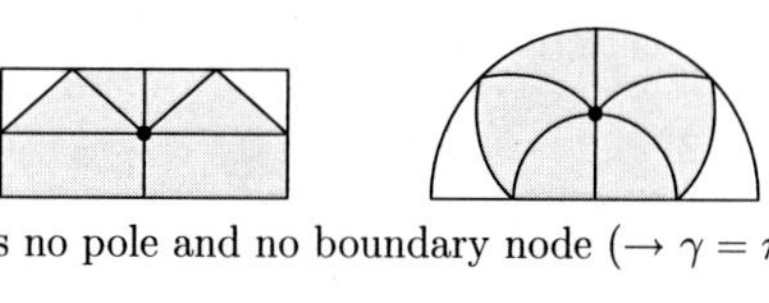

x is no pole and no boundary node ($\rightarrow \gamma = \pi$);

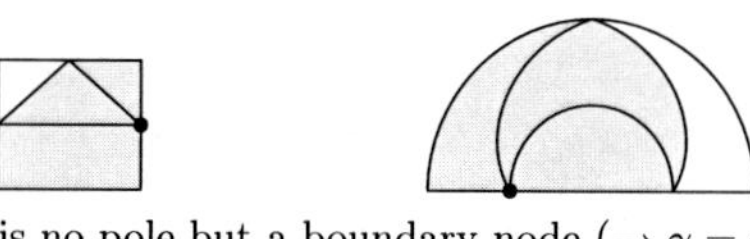

x is no pole but a boundary node ($\rightarrow \gamma = \pi$)

Figure 3.6: Possible structures of ω_{x} with $\vartheta_{-,\omega_{\mathrm{x}}} = 0$ and the corresponding reference domains $\hat{\omega}_{\mathrm{x}}$

Like in the previous proofs, we show Theorem 3.11 for scalar functions, that is, for each component of $\underline{v}$. If the boundary conditions are not distinguished for the single components of $\underline{v}$, then the proof can be written down immediately for vector functions. Let v be a scalar function in $\mathcal{H}^1(\tilde{\omega}_T)$ that vanishes on Γ_D. Recall

$$\mathrm{I}_h\, v(\underline{\mathrm{x}}) = \sum_{\mathrm{x}\in\mathcal{N}_h\setminus\mathcal{N}_{h,D}} (\pi_{0,\omega_\mathrm{x}} v)\, \phi_\mathrm{x}(\underline{\mathrm{x}}) \quad \text{and} \quad \sum_{\mathrm{x}\in\mathcal{N}(T)} \phi_\mathrm{x} \equiv 1.$$

Let $T \in \mathcal{T}_h$ be arbitrary. Note that $v|_T = \sum_{\mathrm{x}\in\mathcal{N}(T)} v|_T\phi_\mathrm{x}$ due to (3.11), see page 45. Obviously,

$$\begin{aligned}(v - \mathrm{I}_h\, v)|_T &= v|_T - \sum_{\mathrm{x}\in\mathcal{N}(T)} (\pi_{0,\omega_\mathrm{x}} v)\, \phi_\mathrm{x} + \sum_{\mathrm{x}\in\mathcal{N}(T)\cap\mathcal{N}_{h,D}} (\pi_{0,\omega_\mathrm{x}} v)\, \phi_\mathrm{x} \\ &= \sum_{\mathrm{x}\in\mathcal{N}(T)} (v - \pi_{0,\omega_\mathrm{x}} v)|_T\, \phi_\mathrm{x} + \sum_{\mathrm{x}\in\mathcal{N}(T)\cap\mathcal{N}_{h,D}} (\pi_{0,\omega_\mathrm{x}} v)\, \phi_\mathrm{x}.\end{aligned}$$

We conclude from (3.11), from the triangle inequality and from Lemma 3.13 that

$$\begin{aligned}\|\lceil v - \mathrm{I}_h\, v\rceil\|_{0,T} &\leq \sum_{\mathrm{x}\in\mathcal{N}(T)} \|\lceil v - \pi_{0,\omega_\mathrm{x}} v\rceil\|_{0,T} + \sum_{\mathrm{x}\in\mathcal{N}(T)\cap\mathcal{N}_{h,D}} \|\lceil(\pi_{0,\omega_\mathrm{x}} v)\, \phi_\mathrm{x}\rceil\|_{0,T} \\ &\leq \sum_{\mathrm{x}\in\mathcal{N}(T)} h_{+,T}\lceil v\rceil_{1,\omega_\mathrm{x}} + \lceil T\rceil^{1/2} \sum_{\mathrm{x}\in\mathcal{N}(T)\cap\mathcal{N}_{h,D}} |\pi_{0,\omega_\mathrm{x}} v|. \end{aligned} \tag{3.43}$$

To estimate the last sum, we exploit that for each $\mathrm{x} \in \mathcal{N}(T)\cap\mathcal{N}_{h,D}$, there is an edge $E' \subset \Gamma_D$ and an element $T' \subset \omega_\mathrm{x}$ with $E' \in \mathcal{E}(T')$ and $v \equiv 0$ on E'. Note that $\lceil v - \pi_{0,\omega_\mathrm{x}} v\rceil_{1,T'} = \lceil v\rceil_{1,T'}$, since $\pi_{0,\omega_\mathrm{x}} v$ is constant over ω_x. The trace theorem (Lemma 3.12) and Lemma 3.13 imply that

$$\begin{aligned}|\pi_{0,\omega_\mathrm{x}} v| &= \lceil E'\rceil^{-1/2}\|\lceil \pi_{0,\omega_\mathrm{x}} v\rceil\|_{0,E'} = \lceil E'\rceil^{-1/2}\|\lceil v - \pi_{0,\omega_\mathrm{x}} v\rceil\|_{0,E'} \\ &\lesssim \lceil T'\rceil^{-1/2}\Big(\|\lceil v - \pi_{0,\omega_\mathrm{x}} v\rceil\|_{0,T'}^2 + h_{+,T'}^2\lceil v - \pi_{0,\omega_\mathrm{x}} v\rceil_{1,T'}^2\Big)^{1/2} \\ &\lesssim \lceil T'\rceil^{-1/2}\Big(h_{+,T'}^2\lceil v\rceil_{1,\omega_\mathrm{x}}^2 + h_{+,T'}^2\lceil v\rceil_{1,T'}^2\Big)^{1/2}.\end{aligned}$$

The size of adjacent elements does not change rapidly so that $\lceil T'\rceil \sim \lceil T\rceil$ and $h_{+,T'} \sim h_{+,T}$. The combination with the estimate (3.43) proves the first assertion of Theorem 3.11.

For edges $E \in \mathcal{E}_h$, we proceed in the same way. Note that $\sum_{\mathrm{x}\in\mathcal{N}(E)} \phi_\mathrm{x}(\underline{\mathrm{x}}) = 1$ for all $\underline{\mathrm{x}} \in E$, since $\phi_\mathrm{x} \equiv 0$ on E for all $\mathrm{x} \notin \mathcal{N}(E)$. Hence,

$$\begin{aligned}v - \mathrm{I}_h\, v|_E &= \sum_{\mathrm{x}\in\mathcal{N}(E)} (v - \pi_{0,\omega_\mathrm{x}} v)|_E\phi_\mathrm{x} + \sum_{\mathrm{x}\in\mathcal{N}(E)\cap\mathcal{N}_{h,D}} (\pi_{0,\omega_\mathrm{x}} v)\phi_\mathrm{x}, \\ \|\lceil v - \mathrm{I}_h\, v\rceil\|_{0,E} &\leq \sum_{\mathrm{x}\in\mathcal{N}(E)} \|\lceil v - \pi_{0,\omega_\mathrm{x}} v\rceil\|_{0,E} + \sum_{\mathrm{x}\in\mathcal{N}(E)\cap\mathcal{N}_{h,D}} \|\lceil \pi_{0,\omega_\mathrm{x}} v\rceil\|_{0,E}.\end{aligned}$$

Now let $T' \in \omega_E$. If $\mathcal{N}(E)\cap\mathcal{N}_{h,D} \neq \emptyset$, let $T' \in \omega_E$ be an element at the Dirichlet boundary with an edge $E' \subset \Gamma_D$. Obviously,

$$\|\lceil \pi_{0,\omega_\mathrm{x}} v\rceil\|_{0,E} = \lceil E\rceil^{1/2}\lceil E'\rceil^{-1/2}\|\lceil \pi_{0,\omega_\mathrm{x}} v\rceil\|_{0,E'} = \lceil E\rceil^{1/2}\lceil E'\rceil^{-1/2}\|\lceil v - \pi_{0,\omega_\mathrm{x}} v\rceil\|_{0,E'}$$

for $\mathrm{x} \in \mathcal{N}(E) \cap \mathcal{N}_{h,D}$.

We use again that $\lceil v - \pi_{0,\omega_\mathrm{x}} v \rceil_{1,T'} = \lceil v \rceil_{1,T'}$, because $\pi_{0,\omega_\mathrm{x}} v$ is constant on T'. Moreover, the size of adjacent elements does not change rapidly, so that the trace theorem (Lemma 3.12) and Lemma 3.13 imply that

$$\begin{aligned} \|\lceil v - \mathrm{I}_h v \rceil\|_{0,E} &\leq \sum_{\mathrm{x} \in \mathcal{N}(E)} \|\lceil v - \pi_{0,\omega_\mathrm{x}} v \rceil\|_{0,E} + \sum_{\mathrm{x} \in \mathcal{N}(E) \cap \mathcal{N}_{h,D}} \frac{\lceil E \rceil^{1/2}}{\lceil E' \rceil^{1/2}} \|\lceil v - \pi_{0,\omega_\mathrm{x}} v \rceil\|_{0,E'} \\ &\lesssim \sum_{\mathrm{x} \in \mathcal{N}(E)} \frac{\lceil E \rceil^{1/2}}{\lceil T' \rceil^{1/2}} \left(\|\lceil v - \pi_{0,\omega_\mathrm{x}} v \rceil\|_{0,T'}^2 + h_{+,T'}^2 \lceil v - \pi_{0,\omega_\mathrm{x}} v \rceil_{1,T'}^2 \right)^{1/2} \\ &\lesssim \frac{\lceil E \rceil^{1/2}}{\lceil T \rceil^{1/2}} h_{+,T} \lceil v \rceil_{1,\tilde{\omega}_E}, \end{aligned}$$

which completes the proof. □

4 Model problems

In this chapter, we discuss two examples, the Laplace and the linear elasticity problems near the vertex of a cone. We will show how the associated eigenvalue problems of the form (1.4) can be derived and explain what the results of Chapter 3 mean regarding these examples. Finally, we present the results of numerical experiments concerning these two model problems.

4.1 Preparations

The mentioned eigenvalue problems are defined on the unit sphere. We will derive associated operator pencils $\mathcal{B}(\cdot) : \mathbb{K} \to \mathcal{L}(V, V)$ in order to apply the theory of Section 2.3. The following lemma gives a necessary and sufficient condition, when an operator is uniquely defined by an inner product and a sesquilinear form.

Lemma 4.1. *Let $\mathbb{K} = \mathbb{R}$ or $\mathbb{K} = \mathbb{C}$ and let V be a Hilbert space over $\mathbb{K}$ with the inner product $(\cdot, \cdot)_V$. Furthermore, let $a : V \times V \to \mathbb{K}$ be a sesquilinear form. An operator $\mathcal{A} \in \mathcal{L}(V, V)$ which is given via the relation*

$$(\mathcal{A}\underline{u}, \underline{v})_V = a(\underline{u}, \underline{v}) \qquad \forall \underline{u}, \underline{v} \in V$$

is well-defined if and only if the sesquilinear form $a(\cdot, \cdot)$ is bounded with respect to the second argument, that is,

$$|a(\underline{u}, \underline{v})| \leq c_{\underline{u}} \|\underline{v}\|_V \qquad \forall \underline{v} \in V$$

for each $\underline{u} \in V$ and some constant $c_{\underline{u}} > 0$.

Proof. For each element $\underline{u} \in V$, the sesquilinear form $a(\cdot, \cdot)$ defines a linear functional $f_{a,\underline{u}}$ via $f_{a,\underline{u}}(\overline{\underline{v}}) = a(\underline{u}, \underline{v})$. If $a(\cdot, \cdot)$ is bounded, then $f_{a,\underline{u}} \in V^\star$. The Riesz representation theorem yields that there is a unique element $\underline{u}_* \in V$ such that $(\underline{u}_*, \underline{v})_V = a(\underline{u}, \underline{v})$ for all $\underline{v} \in V$, see page 13. We set $\mathcal{A}\underline{u} := \underline{u}_*$. Thus, $\mathcal{A}$ is well-defined by $a(\cdot, \cdot)$.

To prove the other direction, suppose that $\mathcal{A}$ is well-defined. Let $\underline{u} \in V$, $\underline{u}_* := \mathcal{A}\underline{u}$ and let $f_{a,\underline{u}}$ be the linear functional which is associated with $a(\cdot, \cdot)$ and $\underline{u}$ via $f_{a,\underline{u}}(\overline{\underline{v}}) = a(\underline{u}, \underline{v})$. The inner product $(\underline{u}_*, \cdot)_V$ defines a linear, *bounded* functional $g_{\mathcal{A}}(\overline{\underline{v}}) = (\underline{u}_*, \underline{v})_V$. If $(\mathcal{A}\underline{u}, \underline{v})_V = a(\underline{u}, \underline{v})$ is satisfied, then $g_{\mathcal{A}}(\overline{\underline{v}}) = f_{a,\underline{u}}(\overline{\underline{v}})$. Thus, $f_{a,\underline{u}}$ is bounded as well, which completes the proof. □

The error estimates derived in Sections 3.4, 3.5 and 3.6 are based on the introduction of element and edge residuals. To compute the edge residuals in the representation (3.12) on page 46, we have to define the jump of a function over an edge E. Let Ω_S be a domain on the unit sphere and let $\mathcal{T}_h$ be a triangulation of Ω_S, see Section 3.1. For each $T \in \mathcal{T}_h$ and $E \in \mathcal{E}(T)$, let $\mathbf{n}_{T,E} = \mathbf{n}_{T,E}(\underline{\mathbf{x}})$ denote the normal vector exterior to T at the point $\underline{\mathbf{x}} \in E$. In addition, we associate with each edge $E \in \mathcal{E}_h$ a normal vector $\mathbf{n}_E$ which points to the exterior of Ω_S if E is a boundary edge and which equals $\mathbf{n}_{T,E}$ for *one* adjacent element $T \in \omega_E$ otherwise (and $-\mathbf{n}_{T',E}$ for $T' \in \omega_E \setminus T$).

The jump $\lfloor \underline{\psi}(\underline{\mathbf{x}}) \rfloor_E$ of a (scalar or vector) function $\underline{\psi}$ over the edge E into the direction of $\mathbf{n}_E$ is defined for $E \in \mathcal{E}_{h,\Omega_S}$ by

$$\lfloor \underline{\psi}(\underline{\mathbf{x}}) \rfloor_E = \lim_{t \to 0+} \underline{\psi}(\underline{\mathbf{x}} + t\,\mathbf{n}_E) - \lim_{t \to 0+} \underline{\psi}(\underline{\mathbf{x}} - t\,\mathbf{n}_E) \qquad \text{a.e. on } E.$$

When we consider functions $\underline{\psi}$ which are defined elementwise, they might be discontinuous. Thus, the definition of the corresponding integral over an edge $E \in \mathcal{E}_{h,\Omega_S}$ is not necessarily evident. Usually, a trace operator is used for a proper introduction of the line integral. For our purposes and for the sake of simplicity, the following definition shall suffice: Let $E \in \mathcal{E}_h$ and $T \in \omega_E$, and let $\underline{v}$ be a (scalar of vector) function which is defined on E in the trace sense so that $\underline{\bar{v}} \cdot \underline{\psi}$ is an integrable scalar function over E in the following sense:

$$\begin{aligned}
\int_E \underline{\bar{v}}(\underline{\mathbf{x}}) \cdot \underline{\psi}(\underline{\mathbf{x}})|_T \,\mathrm{d}\sigma &:= \int_E \underline{\bar{v}}(\underline{\mathbf{x}}) \cdot \lim_{t \to 0+} \underline{\psi}(\underline{\mathbf{x}} - t\,\mathbf{n}_{T,E}(\underline{\mathbf{x}}))\,\mathrm{d}\sigma \\
&= \int_E \underline{\bar{v}}(\underline{\mathbf{x}}) \cdot \lim_{t \to 0+} \underline{\psi}(\underline{\mathbf{x}} - t\,\delta_{T,E}\mathbf{n}_E(\underline{\mathbf{x}}))\,\mathrm{d}\sigma,
\end{aligned}$$

where

$$\mathbf{n}_{T,E} = \delta_{T,E}\,\mathbf{n}_E \qquad \text{and} \quad \delta_{T,E} = \begin{cases} 1 & \text{if } \mathbf{n}_{T,E} = \mathbf{n}_E \\ -1 & \text{if } \mathbf{n}_{T,E} = -\mathbf{n}_E \end{cases}.$$

Depending on $\underline{\psi}$, let $\underline{\varrho}$ be a vector function or a second order tensor valued function, and let $\underline{\psi}(\underline{\mathbf{x}}) := (\underline{\varrho} \cdot \mathbf{n}_{T,E})(\underline{\mathbf{x}})$. Then,

$$\begin{aligned}
\sum_{T \subset \omega_E} \int_E \underline{\bar{v}} \cdot \underline{\varrho} \cdot \mathbf{n}_{T,E}\,\mathrm{d}\sigma &= \sum_{T \subset \omega_E} \int_E \underline{\bar{v}} \cdot \underline{\varrho} \cdot \delta_{T,E}\mathbf{n}_E\,\mathrm{d}\sigma \\
&= -1 \cdot \int_E \underline{\bar{v}}(\underline{\mathbf{x}}) \lim_{t \to 0+} \Big(\underline{\varrho} \cdot \mathbf{n}_E\Big)(\underline{\mathbf{x}} + t\mathbf{n}_E)\,\mathrm{d}\sigma + 1 \cdot \int_E \underline{\bar{v}}(\underline{\mathbf{x}}) \lim_{t \to 0+} \Big(\underline{\varrho} \cdot \mathbf{n}_E\Big)(\underline{\mathbf{x}} - t\mathbf{n}_E)\,\mathrm{d}\sigma \\
&= -\int_E \underline{\bar{v}} \cdot \lfloor \underline{\varrho} \cdot \mathbf{n}_E \rfloor_E\,\mathrm{d}\sigma. \qquad (4.1)
\end{aligned}$$

4.2 The Laplace equation

4.2.1 Error Estimates

Let Ω be an infinite conical domain as described in Chapter 1, see Figure 1.1, and let $\Omega_S := \Omega \cap \mathcal{S}^2$ be its intersection with the unit sphere; this means that

$$\Omega = \{\underline{\mathbf{X}} \in \mathbb{R}^3 \mid \underline{\mathbf{x}} := \underline{\mathbf{X}}/|\underline{\mathbf{X}}| \in \Omega_S,\, 0 < r := |\underline{\mathbf{X}}| < \infty\}, \quad \partial\Omega_S = \partial\Omega \cap \mathcal{S}^2.$$

We consider the Laplace problem

$$\begin{aligned}
-\Delta U &= 0 \quad \text{in } \Omega, \\
U &= 0 \quad \text{on } \partial\Omega_D, \qquad (4.2) \\
\partial U/\partial \mathbf{n} &= 0 \quad \text{on } \partial\Omega_N,
\end{aligned}$$

where $\partial\Omega = \partial\Omega_D \cup \partial\Omega_N$ and $\partial\Omega_D \cap \partial\Omega_N = \emptyset$, compare problem (1.2). Note that $U = 0$ is not the only solution, since Ω is an infinite domain.

After the application of the divergence theorem, the variational formulation of problem (4.2) reads

$$\int_\Omega \nabla U \cdot \nabla V = 0. \tag{4.3}$$

We use the approach (1.3) for U and a similar approach for the test functions V,

$$U(\underline{\mathbf{X}}) = r^\alpha u(\underline{\mathbf{x}}), \qquad V(\underline{\mathbf{X}}) = \Phi(r)v(\underline{\mathbf{x}}), \qquad \underline{\mathbf{X}} \in \Omega,\ r = |\underline{\mathbf{X}}|,\ \underline{\mathbf{x}} = \underline{\mathbf{X}}/|\underline{\mathbf{X}}| \in \Omega_S, \tag{4.4}$$

where $\Phi \in \mathcal{C}_0^\infty(\mathbb{R}_+)$ is a differentiable function with bounded support on $\mathbb{R}_+$. Partial integration yields

$$\int_0^\infty r^{\alpha+1}\Phi'(r)\,\mathrm{d}r = -(\alpha+1)\int_0^\infty r^\alpha \Phi(r)\,\mathrm{d}r. \tag{4.5}$$

Using the notation of Section 2.4, one readily verifies that

$$\nabla U = r^{\alpha-1}\nabla_{\!S} u + \alpha r^{\alpha-1} u\underline{\mathbf{x}}, \qquad \nabla V = \Phi(r) r^{-1}\nabla_{\!S} v + \Phi'(r) v\underline{\mathbf{x}}, \tag{4.6}$$

cf. [75]. With these relations, we can prove the following integral transformations and immediately conclude a special form of the divergence theorem for functions on the sphere.

Lemma 4.2. *The following relations hold:*

$$\begin{aligned}
\int_\Omega \Delta U\, V\,\mathrm{d}\Omega &= \Big(\int_0^\infty r^\alpha \Phi(r)\,\mathrm{d}r\Big)\cdot\Big(\int_{\Omega_S} \Delta_S u\, v + \alpha(\alpha+1) uv\,\mathrm{d}\mathcal{S}\Big),\\
\int_\Omega \nabla U \cdot \nabla V\,\mathrm{d}\Omega &= \Big(\int_0^\infty r^\alpha \Phi(r)\Big)\cdot\Big(\int_{\Omega_S} \nabla_{\!S} u \cdot \nabla_{\!S} v - \alpha(\alpha+1) uv\,\mathrm{d}\mathcal{S}\Big),\\
\int_{\partial\Omega} \{\nabla U \cdot \mathbf{n}\} V\,\mathrm{d}\Sigma &= \Big(\int_0^\infty r^\alpha \Phi(r)\,\mathrm{d}r\Big)\cdot\Big(\int_{\partial\Omega_S} \{\nabla_{\!S} u \cdot \mathbf{n}_S\}\, v\,\mathrm{d}\sigma\Big).
\end{aligned}$$

Proof. According to the definition of $\nabla_{\!S}$, $\underline{\mathbf{x}}$ and $\mathbf{n}_S$ (see pages 31 and 2.4), the relations

$$\nabla_{\!S} u \cdot \underline{\mathbf{x}} = \underline{\mathbf{x}} \cdot \nabla_{\!S} u = 0, \quad \underline{\mathbf{x}} \cdot \underline{\mathbf{x}} = 1, \quad \frac{\partial}{\partial r}\nabla_{\!S} u = \frac{\partial}{\partial r}\underline{\mathbf{x}} = 0 \quad \text{and} \quad \underline{\mathbf{x}} \cdot \mathbf{n}_S = 0$$

hold. Moreover,

$$\nabla_{\!S} \cdot (u\underline{\mathbf{x}}) = \nabla_{\!S} u \cdot \underline{\mathbf{x}} + u\nabla_{\!S} \cdot \underline{\mathbf{x}} = 0 + \sum_{i=1}^2 u\mathbf{g}_S^i \cdot \frac{\partial}{\partial \xi_i}\frac{\partial \underline{\mathbf{X}}}{\partial r} = \sum_{i=1}^2 u\mathbf{g}_S^i \cdot \frac{\partial \mathbf{g}_i}{\partial r} = \sum_{i=1}^2 u\mathbf{g}_S^i \mathbf{g}_i^S = 2u.$$

We exploit that $\Omega = \{\underline{\mathbf{X}} \in \mathbb{R}^3 \mid \underline{\mathbf{X}} = r\underline{\mathbf{x}},\ \underline{\mathbf{x}} \in \Omega_S,\ 0 < r = |\underline{\mathbf{X}}| < \infty\}$ and $\partial\Omega = \{\underline{\mathbf{X}} \in \mathbb{R}^3 \mid \underline{\mathbf{X}} = r\underline{\mathbf{x}},\ \underline{\mathbf{x}} \in \partial\Omega_S,\ 0 < r = |\underline{\mathbf{X}}| < \infty\}$ and conclude from (4.4), (4.5) and (4.6) that

$$\begin{aligned}
\int_\Omega \Delta U\, V\,\mathrm{d}\Omega &= \int_\Omega \{\nabla \cdot \nabla U\}\, V\,\mathrm{d}\Omega\\
&= \int_0^\infty\!\!\int_{\Omega_S} \{(r^{-1}\nabla_{\!S} + \underline{\mathbf{x}}\frac{\partial}{\partial r}) \cdot (r^{\alpha-1}\nabla_{\!S} u + \alpha r^{\alpha-1} u\underline{\mathbf{x}})\}\, \Phi(r) v\, r^2\,\mathrm{d}r\,\mathrm{d}\mathcal{S}\\
&= \int_0^\infty\!\!\int_{\Omega_S} r^{\alpha-2} r^2\, \Phi(r)\, \{\nabla_{\!S} \cdot \nabla_{\!S} u + \alpha\nabla_{\!S} \cdot (u\underline{\mathbf{x}})\}\, v\,\mathrm{d}r\,\mathrm{d}\mathcal{S}
\end{aligned}$$

$$
\begin{aligned}
&+\int_0^\infty\int_{\Omega_S}\Phi(r)\,r^2\,\{\underline{\mathbf{x}}(\alpha-1)r^{\alpha-2}\cdot(\nabla_S u+\alpha u\underline{\mathbf{x}})+r^{\alpha-1}\underline{\mathbf{x}}\cdot\frac{\partial}{\partial r}(\nabla_S u+\alpha u\underline{\mathbf{x}})\}v\,\mathrm{d}r\,\mathrm{d}\mathcal{S}\\
&= \Big(\int_0^\infty r^\alpha\Phi(r)\,\mathrm{d}r\Big)\cdot\Big(\int_{\Omega_S}\{\Delta_S u+2\alpha u+\alpha(\alpha-1)u\}\,v\,\mathrm{d}\mathcal{S}\Big)\\
&= \Big(\int_0^\infty r^\alpha\Phi(r)\,\mathrm{d}r\Big)\cdot\Big(\int_{\Omega_S}\Delta_S u\,v+\alpha(\alpha+1)uv\,\mathrm{d}\mathcal{S}\Big),
\end{aligned}
$$

$$
\begin{aligned}
\int_\Omega \nabla U\cdot\nabla V\,\mathrm{d}\Omega &= \int_0^\infty\int_{\Omega_S}(r^{\alpha-1}\nabla_S u+\alpha r^{\alpha-1}u\underline{\mathbf{x}})\cdot(\Phi(r)r^{-1}\nabla_S v+\Phi'(r)v\underline{\mathbf{x}})\,r^2\,\mathrm{d}r\,\mathrm{d}\mathcal{S}\\
&= \int_0^\infty\int_{\Omega_S} r^2\Big(r^{\alpha-2}\Phi(r)\nabla_S u\cdot\nabla_S v+r^{\alpha-1}\Phi'(r)\nabla_S u\cdot(v\underline{\mathbf{x}})\\
&\qquad +\alpha r^{\alpha-2}\Phi(r)\,u\underline{\mathbf{x}}\cdot\nabla_S v+\alpha r^{\alpha-1}\Phi'(r)\,uv\,\underline{\mathbf{x}}\cdot\underline{\mathbf{x}}\Big)\,\mathrm{d}r\,\mathrm{d}\mathcal{S}\\
&= \Big(\int_0^\infty r^\alpha\Phi(r)\Big)\cdot\Big(\int_{\Omega_S}\nabla_S u\cdot\nabla_S v\,\mathrm{d}\mathcal{S}\Big)+\Big(\int_0^\infty r^{\alpha+1}\Phi'(r)\Big)\cdot\Big(\int_{\Omega_S}\alpha uv\,\mathrm{d}\mathcal{S}\Big)\\
&= \Big(\int_0^\infty r^\alpha\Phi(r)\Big)\cdot\Big(\int_{\Omega_S}\nabla_S u\cdot\nabla_S v-\alpha(\alpha+1)uv\,\mathrm{d}\mathcal{S}\Big),
\end{aligned}
$$

$$
\begin{aligned}
\int_{\partial\Omega}\{\nabla U\cdot\mathbf{n}\}V\,\mathrm{d}\Sigma &= \int_0^\infty\int_{\partial\Omega_S}\{(r^{\alpha-1}\nabla_S u+\alpha r^{\alpha-1}u\underline{\mathbf{x}})\cdot\mathbf{n}_S\}\,\Phi(r)v\,r\,\mathrm{d}r\,\mathrm{d}\sigma\\
&= \Big(\int_0^\infty r^\alpha\Phi(r)\,\mathrm{d}r\Big)\cdot\Big(\int_{\partial\Omega_S}\{\nabla_S u\cdot\mathbf{n}_S\}\,v\,\mathrm{d}\sigma\Big).
\end{aligned}
$$

□

Corollary 4.3 (Divergence theorem on the sphere). *Let $\Delta_S := \nabla_S\cdot\nabla_S$ be the Laplace-Beltrami operator. Then, the relation*

$$
\int_{\Omega_S}\nabla_S u\cdot\nabla_S v\,\mathrm{d}\mathcal{S}=\int_{\Omega_S}-\Delta_S u\,v\,\mathrm{d}\mathcal{S}+\int_{\partial\Omega_S}(\nabla_S u\cdot\mathbf{n}_S)v\,\mathrm{d}\sigma
$$

holds for sufficiently smooth functions u, v as introduced in (4.4).

Proof. Partial integration of the integral $\int_\Omega \Delta U\,V\,\mathrm{d}\Omega$ yields

$$
\begin{aligned}
0 &= \int_\Omega\Delta U\,V\,\mathrm{d}\Omega+\int_\Omega\nabla U\cdot\nabla V\,\mathrm{d}\Omega-\int_{\partial\Omega}\{\nabla U\cdot\mathbf{n}\}V\,\mathrm{d}\Sigma\\
&= \Big(\int_0^\infty r^\alpha\Phi(r)\,\mathrm{d}r\Big)\cdot\\
&\quad\cdot\Big(\int_{\Omega_S}\{\Delta_S u\,v+\alpha(\alpha+1)uv+\nabla_S u\cdot\nabla_S v-\alpha(\alpha+1)uv\,\mathrm{d}\mathcal{S}-\int_{\partial\Omega_S}\{\nabla_S u\cdot\mathbf{n}_S\}\,v\}\,\mathrm{d}\sigma\Big)\\
&= \Big(\int_0^\infty r^\alpha\Phi(r)\,\mathrm{d}r\Big)\cdot\Big(\int_{\Omega_S}\{\Delta_S u\,v+\nabla_S u\cdot\nabla_S v\}\,\mathrm{d}\mathcal{S}-\int_{\partial\Omega_S}\{\nabla_S u\cdot\mathbf{n}_S\}\,v\,\mathrm{d}\sigma\Big).
\end{aligned}
$$

The assertion follows from the division by the integral over r. □

We define the space

$$
V=\{v\in\mathcal{H}^1(\Omega_S)\mid v=0\text{ on }\Gamma_D\}.
$$

With the relations in (4.4) and according to Lemma 4.2, the weak formulation (4.3) of problem (4.2) can be rewritten in the form: Find $\alpha \in \mathbb{R}$, $u \in V$ such that

$$\int_{\Omega_S} \nabla_S u \cdot \nabla_S v \, \mathrm{d}S - \alpha(\alpha+1) \int_{\Omega_S} uv \, \mathrm{d}S = 0 \qquad \forall v \in V. \tag{4.7}$$

This problem is symmetric with respect to u and v; thus there are only real eigenvalues α and the corresponding eigenfunctions can be chosen so that they are real as well. The space V is therefore a real vector space. Moreover, the dual problem corresponding to (4.7) does not differ from the primal problem.

Problem (4.7) is the weak formulation of an eigenvalue problem for the Laplace-Beltrami operator Δ_S,

$$-\Delta_S u = \alpha(\alpha+1)u, \tag{4.8}$$

with homogeneous Dirichlet and Neumann boundary conditions, compare Corollary 4.3.

Lemma 4.4. *The spectrum of problem (4.7) possesses no eigenvalues in the interval $-1 < \alpha < 0$.*

Proof. Since $\alpha(\alpha+1) < 0$ for $-1 < \alpha < 0$, the assertion follows from (4.7) and from $\int_{\Omega_S} |\nabla_S u|^2 \, \mathrm{d}S - \alpha(\alpha+1) \int_{\Omega_S} u^2 \, \mathrm{d}S > 0$ for all $u \neq 0$. □

Remark 4.5. *For $\alpha = -1/2$ and $v = u$, the left-hand side of problem (4.7) equals the norm $\|\!\lceil u \rceil\!\|^2_{1,\Omega_S}$ which was introduced in Section 2.4 (page 31). The statement of Lemma 4.4 was formulated in [55], where the strip $-1 < \mathrm{Re}(\alpha) < 0$ is called the energy strip of the eigenvalue problem (4.7).*

Moreover, it follows from arguments of Fredholm theory, that the spectrum of problem (4.7) consists of isolated eigenvalues with finite multiplicities and the only possible accumulation point at infinity.

Furthermore, the spectrum is symmetric with respect to $-1/2$. This means that all eigenvalues appear in pairs $(\alpha, -1-\alpha)$. The eigenspaces corresponding to α and $-1-\alpha$ are identical. This is easily seen from $\alpha(\alpha+1) = (-1-\alpha)((-1-\alpha)+1)$.

Obviously, the substitution $\lambda = \alpha(\alpha+1)$ reduces (4.7) and (4.8) to linear eigenvalue problems for λ and u. The Laplace-Beltrami operator is frequently studied in the literature [36, 76] and finds application, for instance, in mathematical metrology or heat transfer problems on curved surfaces. The eigenvalue problem $-\Delta_S u = \lambda u$ for the Laplace-Beltrami operator regarding spherical finite elements was treated in [102]. For both formulations (the quadratic eigenvalue problem for α and u and the linear eigenvalue problem for λ and u), we will analyze the error estimates obtained in Chapter 3. We use the indices α and λ for operators or functions to indicate which formulation is used.

To discard the trivial solution u to problem (4.7), we require $(u,u)_H = 1$ for some inner product $(\cdot,\cdot)_H$ which is defined over $V \times V$, see the discussion on page 20. Note that $\bar{u} = u$, since V is a real Hilbert space. For instance, in accordance with Verfürth [104], let $(u,v)_H = \int_{\Omega_S} uv \, \mathrm{d}S$, so that $(u,u)_H = \|\!\lceil u \rceil\!\|^2_{0,\Omega_S}$. We define the operators $F_\alpha : \mathbb{R} \times V \to \mathbb{R} \times V^\star$ and $F_\lambda : \mathbb{R} \times V \to \mathbb{R} \times V^\star$ so that

$$\langle F_\alpha([\alpha,u]), [\mu,v] \rangle = \int_{\Omega_S} (\nabla_S u \cdot \nabla_S v - \alpha(\alpha+1)uv) \, \mathrm{d}S - \mu(\|\!\lceil u \rceil\!\|^2_{0,\Omega_S} - 1)$$

$$\langle F_\lambda([\lambda, u]), [\mu, v]\rangle = \int_{\Omega_S} (\nabla_S u \cdot \nabla_S v - \lambda uv)\,\mathrm{d}\mathcal{S} - \mu(\|\lceil u \rceil\|_{0,\Omega_S}^2 - 1)$$

for all $\lambda, \mu \in \mathbb{R}$, $u, v \in V$. Problem (4.7) is equivalent to $F_\alpha([\alpha, u]) = F_\lambda([\lambda, u]) = 0$ (in the sense of an equality in $\mathbb{R} \times V^\star$, compare (2.16)),

Let $\mathcal{T}_h$ be a triangulation of Ω_S as suggested in Sections 3.1 and 3.2 and let V_h be the corresponding finite element space. The element and edge residuals R_T, R_E and R_N in the representation (3.12) of F (see page 46) are obtained from the partial integration of (4.7), see Corollary 4.3, and equation (4.1): If $u_h \in V_h$ with $\|\lceil u_h \rceil\|_{0,\Omega_S} = 1$, then

$$\begin{aligned}
\langle F_\lambda([\lambda_h, u_h]), [\mu, v]\rangle &= \sum_{T\in\mathcal{T}_h} \int_T (\nabla_S u_h \cdot \nabla_S v - \lambda_h u_h v)\,\mathrm{d}\mathcal{S} \\
&= \sum_{T\in\mathcal{T}_h} \int_T (-\Delta_S u_h - \lambda_h u_h) v\,\mathrm{d}\mathcal{S} - \sum_{E\in\mathcal{E}_{h,\Omega_S}} \int_E \lfloor \nabla_S u_h \cdot \mathbf{n}_E \rfloor_E\, v\,\mathrm{d}\sigma \\
&\quad + \sum_{E\in\mathcal{E}_{h,N}} \int_E (\nabla_S u_h \cdot \mathbf{n}_E) v\,\mathrm{d}\sigma
\end{aligned}$$

and $\langle F_\alpha([\alpha_h, u_h]), [\mu, v]\rangle = \langle F_\lambda([\alpha_h(\alpha_h+1), u_h]), [\mu, v]\rangle$ for all $\mu \in \mathbb{R}$, $v \in V$. In particular, we obtain the residuals

$$\begin{aligned}
R_T &:= -\Delta_S u_h - \lambda_h u_h = -\Delta_S u_h - \alpha_h(\alpha_h+1)u_h, \\
R_E &:= \lfloor \nabla_S u_h \cdot \mathbf{n}_E \rfloor_E, \qquad R_N := \nabla_S u_h \cdot \mathbf{n}_E.
\end{aligned}$$

Let $\tilde{\eta}_T$, η_T and ε_T be defined as in (3.13), (3.15) and (3.16), see page 47f. In Sections 3.4 and 3.5, we proved the relations

$$\|F_\lambda([\lambda_h, u_h])\|_{\mathbb{R}\times V^\star} \lesssim \Big\{\sum_{T\in\mathcal{T}_h} \tilde{\eta}_T^2\Big\}^{1/2} \le \Big\{\sum_{T\in\mathcal{T}_h} \eta_T^2\Big\}^{1/2} + \Big\{\sum_{T\in\mathcal{T}_h} \varepsilon_T^2\Big\}^{1/2} \tag{4.9}$$

and

$$\Big\{\sum_{T\in\mathcal{T}_h} \eta_T^2\Big\}^{1/2} \lesssim \|F_\lambda([\lambda_h, u_h])\|_{\mathbb{R}\times V^\star} + \Big\{\sum_{T\in\mathcal{T}_h} \varepsilon_T^2\Big\}^{1/2} \tag{4.10}$$

(and appropriate estimates for $F_\alpha(\alpha_h, u_h)$), compare Lemmas 3.15, 3.21, 3.22 and their combination with Lemma 2.5. To obtain the final estimates for the errors $\|[\alpha_0, u_0] - [\alpha_h, u_h]\|_{\mathbb{R}\times V}$ and $\|[\lambda_0, u_0] - [\lambda_h, u_h]\|_{\mathbb{R}\times V}$ (see Theorems 3.17, 3.23 and 3.27), we have to show that the assumptions of Lemma 2.4 (page 16) are satisfied. To this end, we define an inner product $(\cdot,\cdot)_V$ on V so that the space V is complete with respect to the induced norm $\|u\|_V = \sqrt{(u,u)_V}$ and so that the theory of Section 2.3 can be applied; for instance, $(u,v)_V = \int_{\Omega_S} uv\,\mathrm{d}\mathcal{S} + \int_{\Omega_S} \nabla_S u \cdot \nabla_S v\,\mathrm{d}\mathcal{S}$ or, in accordance with the definition of the $\mathcal{H}^1$-norm (see Section 2.4, page 31), $(u,v)_V = \frac{1}{4}\int_{\Omega_S} uv\,\mathrm{d}\mathcal{S} + \int_{\Omega_S} \nabla_S u \cdot \nabla_S v\,\mathrm{d}\mathcal{S}$. If Γ_D has a positive measure, then $(u,v)_V = \int_{\Omega_S} \nabla_S u \cdot \nabla_S v\,\mathrm{d}\mathcal{S}$ suffices. Other definitions are also conceivable and might be handier in certain applications, see Section 4.3.

We define the space

$$H := \mathcal{H}^0(\Omega_S)$$

and the sesquilinear forms $k : V \times V \to \mathbb{R}$, $m : H \times H \to \mathbb{R}$,

$$k(u,v) := (u,v)_V \qquad \text{and} \qquad m(u,v) := \int_{\Omega_S} uv\,\mathrm{d}\mathcal{S}.$$

For each of the suggested inner products, the forms k and m are bounded with respect to $\|v\|_V$. In the case of homogeneous Dirichlet boundary (with positive measure of Γ_D) and $(u,v)_V = \int_{\Omega_S} \nabla_S u \cdot \nabla_S v \, \mathrm{d}\mathcal{S}$, the constant in the relation $m(u,v) \lesssim \|v\|_V$ depends on $\lceil u \rceil_{0,\Omega_S}$ and on the constants from the Friedrichs-type inequality (compare Lemma 3.14 on page 46). According to Lemma 4.1, there are operators $\mathcal{K}, \mathcal{M} : V \to V$ such that

$$(\mathcal{K}u, v)_V = k(u,v), \quad (\mathcal{M}u, v)_V = m(u,v) \qquad \forall u, v \in V. \tag{4.11}$$

Obviously, $\mathcal{K}$ is the identity operator on V. Moreover, the operator $\mathcal{M}$ is compact, since the form m is defined on $H \times H$ and the space V is compactly embedded into H, see [88, Section 3.3].

If $\Gamma_D \neq \emptyset$ and $(u,v)_V = \int_{\Omega_S} \nabla_S u \cdot \nabla_S v \, \mathrm{d}\mathcal{S}$, the operator pencil $\mathcal{B}_\alpha : \mathbb{R} \to \mathcal{L}(V,V)$ which is associated with the eigenvalue problem (4.7) is given by $\mathcal{B}_\alpha(\alpha) = \mathcal{K} - \alpha\mathcal{M} - \alpha^2\mathcal{M}$, compare the definition of F_α and formula (2.15) on page 21. Likewise, for $\lambda = \alpha(\alpha+1)$, the appropriate operator pencil is given by $\mathcal{B}_\lambda : \mathbb{R} \to \mathcal{L}(V,V)$, $\mathcal{B}_\lambda(\lambda) = \mathcal{K} - \lambda\mathcal{M}$. Note that $\mathcal{B}_\alpha(\cdot)$ is *quadratic* in α, whereas $\mathcal{B}_\lambda(\cdot)$ is a *linear* polynomial in λ. Furthermore, $(\mathcal{B}_\lambda'(\lambda)u, u)_V = -m(u,u) = -\lceil u \rceil_{0,\Omega_S} \neq 0$ for all $u \neq 0$, and $(\mathcal{B}_\alpha'(\alpha)u, u)_V = -m(u,u) - 2\alpha m(u,u) = -(1+2\alpha)\lceil u \rceil_{0,\Omega_S} \neq 0$ for all $u \neq 0$ if and only if $\alpha \neq -1/2$. According to Lemma 4.4, the value $\alpha = -1/2$ is not an eigenvalue of problem (4.7). By Lemma 2.3, the pencils $\mathcal{B}_\alpha(\cdot)$ and $\mathcal{B}_\lambda(\cdot)$ consist of Fredholm operators with the index zero so that the assumption (2.17) of Theorem 2.8 (see page 22) is satisfied. Due to Remark 2.11 and Lemma 2.10, the assumptions of Lemma 2.4 are satisfied as well. Consequently,

$$|\alpha_0 - \alpha_h| + \lceil u_0 - u_h \rceil_{1,\Omega_S} = \|[\alpha_0, u_0] - [\alpha_h, u_h]\|_{\mathbb{R}\times V} \sim \|F_\alpha([\alpha_h, u_h])\|_{\mathbb{R}\times V^\star}$$

if α_0 is a simple eigenvalue of $F_\alpha([\alpha, u]) = 0$ and $[\alpha_h, u_h]$ is a finite element approximation of $[\alpha_0, u_0]$. Likewise,

$$|\lambda_0 - \lambda_h| + \lceil u_0 - u_h \rceil_{1,\Omega_S} = \|[\lambda_0, u_0] - [\lambda_h, u_h]\|_{\mathbb{R}\times V} \sim \|F_\lambda([\lambda_h, u_h])\|_{\mathbb{R}\times V^\star}$$

if λ_0 is a simple eigenvalue of $F([\lambda, u]) = 0$ and $[\lambda_h, u_h]$ is a finite element approximation of $[\lambda_0, u_0]$.

Since the primal and the dual problems coincide for the eigenvalue problem (4.7), we have $R_T^\star = R_T$, $R_E^\star = R_E$, $\eta_T^\star = \eta_T$ and $\varepsilon_T^\star = \varepsilon_T$. Thus, the assertions of Section 3.6 hold in the form

$$\begin{aligned} |\alpha_0 - \alpha_h| &\lesssim \sum_{T\in\mathcal{T}_h} \tilde{\eta}_T^2 \le \sum_{T\in\mathcal{T}_h} (\eta_T^2 + \varepsilon_T^2), \\ |\lambda_0 - \lambda_h| &\lesssim \sum_{T\in\mathcal{T}_h} \tilde{\eta}_T^2 \le \sum_{T\in\mathcal{T}_h} (\eta_T^2 + \varepsilon_T^2). \end{aligned}$$

Remark 4.6. *The finite element solution has to approximate the exact solution well enough in the sense of Lemma 2.4. According to Remark 2.11, the admissible difference of the eigenpairs $[\alpha_h, u_h]$ from $[\alpha_0, u_0]$ for the quadratic eigenvalue problem is smaller than the admissible difference of the eigenpairs $[\lambda_h, u_h]$ and $[\lambda_0, u_0]$ for the linear eigenvalue problem.*

Moreover, the values of the terms $\tilde{\eta}_T$, η_T and ε_T are independent of the formulation with respect to α or λ as long as $\lambda = \alpha(\alpha+1)$. Note that $\lambda_0 - \lambda_h = \alpha_0(\alpha_0+1) - \alpha_h(\alpha_h+1) = (\alpha_0 - \alpha_h)(\alpha_0 + \alpha_h + 1) \approx (\alpha_0 - \alpha_h)(2\alpha_0 + 1)$. As the strip $-1 < \mathrm{Re}(\alpha) < 0$ is free of eigenvalues (see Lemma 4.4), we conclude that

$$|\lambda_0 - \lambda_h| \ge |\alpha_0 - \alpha_h|.$$

This means that the upper error bounds for the eigenvalues λ of the (linear) eigenvalue problem for the Laplace-Beltrami operator provide immediately upper error bounds for the singular exponents α. Numerical results for the above estimates are presented in Section 4.5.

In the more general case, where the Dirichlet boundary part is not required, we can choose, for example, the inner product $(u,v)_V = \int_{\Omega_S} uv \, \mathrm{d}\mathcal{S} + \int_{\Omega_S} \nabla_{\mathcal{S}} u \cdot \nabla_{\mathcal{S}} v \, \mathrm{d}\mathcal{S}$ (without the factor 1/4). Then, problem (4.7) is equivalent to $\mathcal{B}_\alpha(\alpha)u = 0$ with the Fredholm operator pencil $\mathcal{B}_\alpha(\alpha) = \mathcal{K} - (\alpha(\alpha+1)+1)\mathcal{M}$. Moreover, the assumptions of Theorem 2.8 are satisfied according to Lemma 4.4. Thus, we obtain the same results as before for the errors

$$\|[\alpha_0, u_0] - [\alpha_h, u_h]\|_{\mathbb{R}\times V} = |\alpha_0 - \alpha_h| + \lceil\!\lceil u_0 - u_h \rceil\!\rceil_{0,\Omega_S} + \lceil u_0 - u_h \rceil_{1,\Omega_S}$$

and $|\alpha_0 - \alpha_h|$ with slightly different constants.

The inner product $(u,v)_V = \frac{1}{4}\int_{\Omega_S} uv \, \mathrm{d}\mathcal{S} + \int_{\Omega_S} \nabla_{\mathcal{S}} u \cdot \nabla_{\mathcal{S}} v \, \mathrm{d}\mathcal{S}$ (which is associated with the norm $\lceil\!\lceil\cdot\rceil\!\rceil_{1,\Omega_S}$ defined on page 31) produces the (Fredholm) operator pencil $\mathcal{B}_\alpha(\alpha) = \mathcal{K} - (\alpha(\alpha+1)+\frac{1}{4}) = \mathcal{K} - (\alpha+\frac{1}{2})^2\mathcal{M}$. This form has the advantage that the corresponding discretized eigenvalue problem is equivalent to a standard eigenvalue problem for a Hamiltonian matrix, when we substitute $\lambda = \alpha + 1/2$. (Note that λ is an eigenvalue if and only if $-\lambda$ is an eigenvalue.) This structure can be exploited for an efficient computation of the eigenvalues λ. In the implementation of problem (4.7), we use the discretization of exactly this form. Once an eigenvalue λ is computed, the corresponding singularity exponent α is obtained immediately from $\alpha = \lambda - 1/2$. The above error estimates hold for

$$\|[\alpha_0, u_0] - [\alpha_h, u_h]\|_{\mathbb{R}\times V} = |\alpha_0 - \alpha_h| + \lceil\!\lceil u_0 - u_h \rceil\!\rceil_{1,\Omega_S} = |\lambda_0 - \lambda_h| + \lceil\!\lceil u_0 - u_h \rceil\!\rceil_{1,\Omega_S}.$$

We remark that the residuals and therefore the error estimators are independent of the inner product $(\cdot,\cdot)_V$. The specific choice of the inner product on V influences only the structure of the operator pencils $\mathcal{B}_\lambda(\cdot)$ or $\mathcal{B}_\alpha(\cdot)$ and the constants in the error estimates, see Section 4.2.2.

4.2.2 Constants

Let us shortly discuss the constants in the error estimates for the eigenvalue problem $-\Delta_{\mathcal{S}} u = \lambda u$. According to Section 3.7 and Lemma 2.4, the constant in the estimate

$$\|F_\lambda([\lambda_h, u_h])\|_{\mathbb{R}\times V^\star} \le C_{F,+}\Big\{\sum_{T\in\mathcal{T}_h} \tilde{\eta}_T^2\Big\}^{1/2} \le C_{F,+}\Big(\Big\{\sum_{T\in\mathcal{T}_h} \eta_T^2\Big\}^{1/2} + \Big\{\sum_{T\in\mathcal{T}_h} \varepsilon_T^2\Big\}^{1/2}\Big)$$

(compare (4.9) on page 78) is bounded from above by

$$\max\{\max_{T\in\mathcal{T}_h} C_{\mathrm{I}_h,T}, \max_{E\in\mathcal{E}_h} C_{\mathrm{I}_h,E}\} \cdot (6\mathcal{Z}+3)\, C_{1,V},$$

where $C_{\mathrm{I}_h,T}, C_{\mathrm{I}_h,E}$ are the constants in the interpolation error estimates, $\mathcal{Z}$ is the maximum number of elements with a common vertex and $C_{1,V}$ is the constant in $\lceil u \rceil_{1,\Omega_S} \le C_{1,V}\|u\|_V$. Thus, $C_{1,V} = 1$ for the Dirichlet problem with the norm $\|u\|_V = \lceil u \rceil_{1,\Omega_S}$. Moreover,

$$\|[\lambda_0, u_0] - [\lambda_h, u_h]\|_{\mathbb{R}\times V} \le 2\|\mathrm{D}F([\lambda_0, u_0])^{-1}\|_{\mathcal{L}(\mathbb{K}\times V^\star, \mathbb{R}\times V)} \|F_\lambda([\lambda_h, u_h])\|_{\mathbb{R}\times V^\star}$$

according to Lemma 2.4. We cannot give an analytic bound for $\|DF([\lambda_0, u_0])^{-1}\|_{\mathcal{L}(\mathbb{K}\times V^\star, \mathbb{R}\times V)}$. According to Lemma 2.12, this constant involves the term $|(\mathcal{B}'_\lambda(\lambda_0)u_0, u_0)_V^{-1}|$ which can be computed as follows: Suppose that $\mathcal{B}_\lambda(\lambda) = \mathcal{K} - \lambda\mathcal{M}$ with $\mathcal{M}$, $\mathcal{K}$ introduced in (4.11). Then,

$$|(\mathcal{B}'_\lambda(\lambda_0)u_0, u_0)_V^{-1}| \;=\; |-m(u_0, u_0^\star)^{-1}| = \frac{1}{\left|\int_{\Omega_S} u_0 u_0^\star \,\mathrm{d}\mathcal{S}\right|}.$$

The constant in the lower bound

$$\Big\{\sum_{T\in\mathcal{T}_h} \eta_T^2\Big\}^{1/2} \le C_{err,-}\Big(\|[\lambda_0, u_0] - [\lambda_h, u_h]\|_{\mathbb{R}\times V} + \Big\{\sum_{T\in\mathcal{T}_h} \varepsilon_T^2\Big\}^{1/2}\Big)$$

is bounded from above by

$$\max\{12(\mathcal{Z}+1)C_{V,1}C_{\eta_T,2}\,\|DF([\lambda_0, \underline{u}_0])\|_{\mathcal{L}(\mathbb{K}\times V, \mathbb{K}\times V^\star)}, C_{1,V}(C_F + 3C_{tr}(C_F+1))\},$$

compare Section 3.7. Moreover

$$\|DF([\lambda_0, u_0])\|_{\mathcal{L}(\mathbb{R}\times V, \mathbb{R}\times V^\star)} \le \|\mathcal{B}'_\lambda(\lambda_0)\, u_0\|_V + \|\mathcal{B}_\lambda(\lambda_0)\|_{\mathcal{L}(V,V)} + 2\,\|u_0\|_H$$

according to Lemma 2.12. Note that

$$\begin{aligned} \|\mathcal{M}u_0\|_V^2 \;&=\; (\mathcal{M}u_0, \mathcal{M}u_0)_V \;=\; m(u_0, \mathcal{M}u_0) \;\le\; \|\lceil u_0\rceil\|_{0,\Omega_S}\|\lceil \mathcal{M}u_0\rceil\|_{0,\Omega_S} \\ &\le\; C_{0,V}\|\lceil u_0\rceil\|_{\Omega_S}\|\mathcal{M}u_0\|_V, \end{aligned}$$

where $C_{0,V}$ is the constant in $\|\lceil u\rceil\|_{0,\Omega_S} \le C_{0,V}\|u\|_V$. In particular, $C_{0,V}$ is the constant in the Friedrichs-type inequality for the domain Ω_S if $(u,v)_V = \int_{\Omega_S} \nabla_{\mathcal{S}} u \nabla_{\mathcal{S}} v \,\mathrm{d}\mathcal{S}$ or $C_{0,V} = 1$ if $(u,v)_V = \int_{\Omega_S} \nabla_{\mathcal{S}} u \nabla_{\mathcal{S}} v \,\mathrm{d}\mathcal{S} + \int_{\Omega_S} uv \,\mathrm{d}\mathcal{S}$ or $C_{0,V} = 2$ if $(u,v)_V = \int_{\Omega_S} \nabla_{\mathcal{S}} u \nabla_{\mathcal{S}} v \,\mathrm{d}\mathcal{S} + \frac{1}{4}\int_{\Omega_S} uv \,\mathrm{d}\mathcal{S}$.

Recall that the sesquilinear form $k(\cdot,\cdot)$ was defined so that $\mathcal{K}$ is the identity operator on V. For $\mathcal{B}_\lambda(\lambda) = \mathcal{K} - \lambda\mathcal{M}$, we conclude that

$$\begin{aligned} \|\mathcal{B}'_\lambda(\lambda_0)\, u_0\|_V \;&=\; \|\mathcal{M}u_0\|_V \;\le\; C_{0,V}\|\lceil u_0\rceil\|_{\Omega_S}, \\ \|\mathcal{B}_\lambda(\lambda_0)\|_{\mathcal{L}(V,V)} \;&=\; \sup_{\substack{u\in V \\ \|u\|_V = 1}} \|(\mathcal{K} - \lambda_0\mathcal{M})u\|_V \;\le\; 1 + |\lambda_0| C_{0,V}, \end{aligned}$$

so that the dependencies of the constant $C_{err,-}$ on an exact solution $[\lambda_0, u_0] \in \mathbb{R}\times V$ to the eigenvalue problem (4.7) are explicitly known.

4.3 The linear elasticity problem

4.3.1 Error estimates

Let Ω be an open, connected domain in $\mathbb{R}^3$. Three different formulations of the linear elasticity problem can be found in the engineering literature: the pure displacement model, where stresses and strains are eliminated; the mixed method by Hellinger and Reissner, where the stresses and displacements are kept in the model, while the strains are eliminated; and the mixed method by Hu and Washizu, where all three terms (displacement, stresses, strains) are kept. An overview of these three methods is given, for instance, by Braess [19]. We adopt the pure displacement model. The Lamé equations are given by

$$\begin{aligned} -\operatorname{div}\sigma(\underline{U}) &= \underline{f} \quad \text{in } \Omega, \\ \underline{U} &= 0 \quad \text{on } \partial\Omega_D, \\ \sigma(\underline{U})\cdot\mathbf{n} &= \underline{g} \quad \text{on } \partial\Omega_N, \end{aligned} \tag{4.12}$$

cf. problem (1.1), where the stress tensor $\sigma(\underline{U})$ is related to the strain tensor $\varepsilon(\underline{U})$ by

$$\sigma(\underline{U}) = A : \varepsilon(\underline{U}), \qquad \varepsilon(\underline{U}) := \frac{1}{2}\Big(\nabla\underline{U} + (\nabla\underline{U})^{\top}\Big)$$

with the (real and piecewise constant) fourth order material tensor A. Moreover, $\partial\Omega_D$ and $\partial\Omega_N$ denote the Dirichlet and Neumann boundary of Ω. We remark that there are three degrees of freedom for each point $\underline{\mathbf{X}} \in \Omega$ so that it is possible to define the Dirichlet and Neumann boundary conditions separately for each degree of freedom, this means for each component of the function $\underline{U}$, compare Remark 3.1.

When Ω has angular points, then the solution to problem (4.12) is composed of a singular and a regular part. It was motivated in Chapter 1 that we can assume, without loss of generality, that Ω coincides with an infinite conical domain. Moreover, we are only interested in the singular part of the solution so that a homogeneous right-hand side can be used, cf. problem (1.2). Thus, we search for functions $\underline{U}$ of the form (1.3),

$$\underline{U}(\underline{\mathbf{X}}) = r^{\alpha}\underline{u}(\underline{\mathbf{x}}), \qquad \underline{\mathbf{X}} \in \Omega, \quad r = |\underline{\mathbf{X}}|, \quad \underline{\mathbf{x}} = \underline{\mathbf{X}}/|\underline{\mathbf{X}}|,$$

which satisfy

$$\begin{aligned} \operatorname{div}\sigma(\underline{U}) &= 0 \quad \text{in } \Omega, \\ \underline{U} &= 0 \quad \text{on } \partial\Omega_D, \\ \mathbf{n}\cdot\sigma(\underline{U}) &= 0 \quad \text{on } \partial\Omega_N. \end{aligned} \tag{4.13}$$

Let $\Omega_{\mathcal{S}} := \Omega \cap \mathcal{S}^2$ be the intersection of Ω with the unit sphere; this means that

$$\Omega = \{\underline{\mathbf{X}} \in \mathbb{R}^3 \mid \underline{\mathbf{x}} := \underline{\mathbf{X}}/|\underline{\mathbf{X}}| \in \Omega_{\mathcal{S}},\, 0 < r := |\underline{\mathbf{X}}| < \infty\}, \quad \partial\Omega_{\mathcal{S}} = \partial\Omega \cap \mathcal{S}^2.$$

We denote the Dirichlet and the Neumann boundary of $\Omega_{\mathcal{S}}$ by $\Gamma_D = \partial\Omega_D \cap \mathcal{S}^2$ and $\Gamma_N = \partial\Omega_N \cap \mathcal{S}^2$, respectively.

To derive a variational formulation for problem (4.13), we assume that Ω can be split into disjoint Lipschitz domains so that the divergence theorem

$$\int_{\Omega} \operatorname{div}\underline{\Psi}\,\mathrm{d}\Omega = \int_{\partial\Omega} \underline{\Psi}\cdot\mathbf{n}\,\mathrm{d}\Sigma.$$

holds for (sufficiently smooth) vector functions $\underline{\Psi}$. We choose, in particular, $\underline{\Psi} := \underline{\bar{V}}\cdot\sigma(\underline{U})$ with the vector function $\underline{U}$ and a (complex) test function $\underline{V}$, where $\underline{\bar{V}}$ denotes the conjugate complex of $\underline{V}$. Note that $\operatorname{div}(\underline{\bar{V}}\cdot\sigma(\underline{U})) = \underline{\bar{V}}\cdot\operatorname{div}\sigma(\underline{U}) + \varepsilon(\underline{\bar{V}}) : \sigma(\underline{U})$. Hence, the variational formulation of (4.13) is given by

$$\begin{aligned} 0 &= -\int_{\Omega} \underline{\bar{V}}\cdot\operatorname{div}\sigma(\underline{U}) = \int_{\Omega} \varepsilon(\underline{\bar{V}}) : \sigma(\underline{U})\,\mathrm{d}\mathcal{S} - \int_{\partial\Omega} \underline{\bar{V}}\cdot\sigma(\underline{U})\cdot\mathbf{n}\,\mathrm{d}\Sigma \\ &= \int_{\Omega} \varepsilon(\underline{\bar{V}}) : \sigma(\underline{U})\,\mathrm{d}\mathcal{S} \end{aligned} \tag{4.14}$$

for $\underline{V}$ vanishing on $\partial\Omega_D$. We use complex-valued test functions $\underline{V}$, because the solution $\underline{U}$ to problem (4.13) might be complex-valued as well, see Chapter 1.

Remark 4.7. *As we consider complex-valued vector functions, one could think of introducing the strain tensor as* $\frac{1}{2}(\nabla \underline{U} + (\nabla \underline{U})^H)$*. The* linear *elasticity theory, however, is concerned with the* linear *part of the Cauchy–Green strain tensor. Obviously,* $\nabla \underline{U} + (\nabla \underline{U})^H$ *is not linear with respect to* $\underline{U}$*, whereas* $\nabla \underline{U} + (\nabla \underline{U})^\top$ *is, so that our definition of* $\varepsilon(\underline{U})$ *is reasonable.*

In order to derive a modified form of the formulation in (4.14), where the specific structure of $\underline{U}$ is exploited, we use test functions $\underline{V}$ with a similar structure; let

$$\underline{U}(\underline{\mathbf{X}}) = r^\alpha \underline{u}(\underline{\mathbf{x}}), \qquad \underline{V}(\underline{\mathbf{X}}) = \Phi(r)\underline{v}(\underline{\mathbf{x}}), \tag{4.15}$$

where $\Phi \in C_0^\infty(\mathbb{R}_+)$ is a differentiable function with bounded support on $\mathbb{R}_+$. In particular, relation (4.5) holds. Using the notation of Section 2.4, one will readily verify that

$$\nabla \underline{U} = r^{\alpha-1}\nabla_{\mathcal{S}}\underline{u} + \alpha\, r^{\alpha-1}\underline{\mathbf{x}}\,\underline{u}, \qquad \nabla \underline{V} = \Phi(r)\, r^{-1}\nabla_{\mathcal{S}}\underline{v} + \Phi'(r)\,\underline{\mathbf{x}}\,\underline{v}. \tag{4.16}$$

For abbreviation, we introduce the special second order tensors

$$\varepsilon_{\mathcal{S}}(\underline{u}) := \frac{1}{2}(\nabla_{\mathcal{S}}\underline{u} + (\nabla_{\mathcal{S}}\underline{u})^\top), \quad \varepsilon_3(\underline{u}) := \frac{1}{2}(\underline{\mathbf{x}\underline{u}} + (\underline{\mathbf{x}\underline{u}})^\top) \tag{4.17}$$

and

$$\sigma_{\mathcal{S}}(\alpha, \underline{u}) := A : (\varepsilon_{\mathcal{S}}(\underline{u}) + \alpha\,\varepsilon_3(\underline{u})). \tag{4.18}$$

It follows immediately from (4.16) that the relations

$$\varepsilon(\underline{U}) = r^{\alpha-1}\varepsilon_{\mathcal{S}}(\underline{u}) + \alpha\, r^{\alpha-1}\varepsilon_3(\underline{u}), \quad \varepsilon(\underline{V}) = r^{-1}\Phi(r)\varepsilon_{\mathcal{S}}(\underline{v}) + \Phi'(r)\varepsilon_3(\underline{v}) \tag{4.19}$$

and

$$\sigma(\underline{U}) = A : \varepsilon(\underline{U}) = r^{\alpha-1}\sigma_{\mathcal{S}}(\alpha, \underline{u}) \tag{4.20}$$

hold. Straightforward computation yields the following Lemma.

Lemma 4.8. *The integrals in the weak formulation (4.14) of problem (4.13) transform to*

$$\begin{aligned}
\int_\Omega \bar{\underline{V}} \cdot \operatorname{div}\sigma(\underline{U})\,\mathrm{d}\Omega &= \Big(\int_0^\infty r^\alpha \Phi(r)\,\mathrm{d}r\Big) \cdot \\
&\quad \cdot \Big(\int_{\Omega_{\mathcal{S}}} \bar{\underline{v}} \cdot \{\nabla_{\mathcal{S}} \cdot \sigma_{\mathcal{S}}(\alpha, \underline{u})\}\,\mathrm{d}\mathcal{S} + (\alpha - 1)\int_{\Omega_{\mathcal{S}}} \bar{\underline{v}} \cdot \{\underline{\mathbf{x}} \cdot \sigma_{\mathcal{S}}(\alpha, \underline{u})\}\Big), \\
\int_\Omega \varepsilon(\bar{\underline{V}}) : \sigma(\underline{U})\,\mathrm{d}\Omega &= \Big(\int_0^\infty r^\alpha \Phi(r)\,\mathrm{d}r\Big) \cdot \Big(\int_{\Omega_{\mathcal{S}}} \Big[\varepsilon_{\mathcal{S}}(\bar{\underline{v}}) - (\alpha+1)\,\varepsilon_3(\bar{\underline{v}})\Big] : \sigma_{\mathcal{S}}(\alpha, \underline{u})\,\mathrm{d}\mathcal{S}\Big), \\
\int_{\partial\Omega} \bar{\underline{V}} \cdot \sigma(\underline{U}) \cdot \mathbf{n}\,\mathrm{d}\Sigma &= \Big(\int_0^\infty r^\alpha \Phi(r)\,\mathrm{d}r\Big) \cdot \Big(\int_{\partial\Omega_{\mathcal{S}}} \bar{\underline{v}} \cdot \sigma_{\mathcal{S}}(\alpha, \underline{u}) \cdot \mathbf{n}_{\mathcal{S}}\,\mathrm{d}\sigma\Big).
\end{aligned}$$

Proof. We exploit that $\Omega = \{\underline{\mathbf{X}} \in \mathbb{R}^3 \mid \underline{\mathbf{X}} = r\underline{\mathbf{x}},\ \underline{\mathbf{x}} \in \Omega_{\mathcal{S}},\ 0 < r = |\underline{\mathbf{X}}| < \infty\}$ and $\partial\Omega = \{\underline{\mathbf{X}} \in \mathbb{R}^3 \mid \underline{\mathbf{X}} = r\underline{\mathbf{x}},\ \underline{\mathbf{x}} \in \partial\Omega_{\mathcal{S}},\ 0 < r = |\underline{\mathbf{X}}| < \infty\}$ as well as the relations (4.15), (4.5) and (4.17)–(4.20) to conclude that

$$\begin{aligned}
\int_\Omega \bar{\underline{V}} \cdot \operatorname{div}\sigma(\underline{U})\,\mathrm{d}\Omega &= \int_\Omega \bar{\underline{V}} \cdot \{\nabla \cdot \sigma(\underline{U})\}\,\mathrm{d}\Omega \\
&= \int_0^\infty \int_{\Omega_{\mathcal{S}}} \Phi(r)\,\bar{\underline{v}} \cdot \{(r^{-1}\nabla_{\mathcal{S}} + \underline{\mathbf{x}}\frac{\partial}{\partial r}) \cdot r^{\alpha-1}\sigma_{\mathcal{S}}(\alpha, \underline{u})\}\, r^2\,\mathrm{d}\mathcal{S}\,\mathrm{d}r
\end{aligned}$$

$$
\begin{aligned}
&= \int_0^\infty \int_{\Omega_S} r^{\alpha-2} r^2\, \Phi(r)\, \underline{\bar{v}} \cdot \{\nabla_S \cdot \sigma_S(\alpha, \underline{u})\}\, \mathrm{d}\mathcal{S}\, \mathrm{d}r \\
&\quad + \int_0^\infty \int_{\Omega_S} r^2\, \Phi(r)\, \underline{\bar{v}} \cdot \{\underline{\mathbf{x}} \cdot (\alpha-1)\, r^{\alpha-2}\, \sigma_S(\alpha, \underline{u})\}\, \mathrm{d}\mathcal{S}\, \mathrm{d}r \\
&= \Big(\int_0^\infty r^\alpha \Phi(r)\, \mathrm{d}r\Big) \cdot \Big(\int_{\Omega_S} \underline{\bar{v}} \cdot \{\nabla_S \cdot \sigma_S(\alpha, \underline{u})\}\, \mathrm{d}\mathcal{S} + (\alpha-1) \int_{\Omega_S} \underline{\bar{v}} \cdot \{\underline{\mathbf{x}} \cdot \sigma_S(\alpha, \underline{u})\}\Big),
\end{aligned}
$$

$$
\begin{aligned}
\int_\Omega \varepsilon(\underline{\bar{V}}) : \sigma(\underline{U})\, \mathrm{d}\Omega &= \int_0^\infty \int_{\Omega_S} \Big(r^{-1} \Phi(r) \varepsilon_S(\underline{\bar{v}}) + \Phi'(r) \varepsilon_3(\underline{\bar{v}})\Big) : \Big(r^{\alpha-1} \sigma_S(\alpha, \underline{u})\Big)\, r^2\, \mathrm{d}\mathcal{S}\, \mathrm{d}r \\
&= \Big(\int_0^\infty r^{\alpha-2}\, r^2\, \Phi(r)\, \mathrm{d}r\Big) \cdot \Big(\int_{\Omega_S} \varepsilon_S(\underline{\bar{v}}) : \sigma_S(\alpha, \underline{u})\, \mathrm{d}\mathcal{S}\Big) \\
&\quad + \Big(\int_0^\infty r^{\alpha-1} r^2\, \Phi'(r)\, \mathrm{d}r\Big) \cdot \Big(\int_{\Omega_S} \varepsilon_3(\underline{\bar{v}}) : \sigma_S(\alpha, \underline{u})\, \mathrm{d}\mathcal{S}\Big) \\
&= \Big(\int_0^\infty r^\alpha\, \Phi(r)\, \mathrm{d}r\Big) \cdot \Big(\int_{\Omega_S} \varepsilon_S(\underline{\bar{v}}) : \sigma_S(\alpha, \underline{u})\, \mathrm{d}\mathcal{S}\Big) \\
&\quad - (\alpha+1) \Big(\int_0^\infty r^\alpha\, \Phi(r)\, \mathrm{d}r\Big) \cdot \Big(\int_{\Omega_S} \varepsilon_3(\underline{\bar{v}}) : \sigma_S(\alpha, \underline{u})\, \mathrm{d}\mathcal{S}\Big) \\
&= \Big(\int_0^\infty r^\alpha \Phi(r)\, \mathrm{d}r\Big) \cdot \Big(\int_{\Omega_S} \Big[\varepsilon_S(\underline{\bar{v}}) - (\alpha+1)\, \varepsilon_3(\underline{\bar{v}})\Big] : \sigma_S(\alpha, \underline{u})\, \mathrm{d}\mathcal{S}\Big),
\end{aligned}
$$

$$
\begin{aligned}
\int_{\partial\Omega} \underline{\bar{V}} \cdot \sigma(\underline{U}) \cdot \mathbf{n}\, \mathrm{d}\Sigma &= \int_0^\infty \int_{\partial\Omega_S} \Phi(r)\, \underline{\bar{v}} \cdot r^{\alpha-1} \sigma_S(\alpha, \underline{u}) \cdot \mathbf{n}_S\, r\, \mathrm{d}\sigma\, \mathrm{d}r \\
&= \Big(\int_0^\infty r^\alpha\, \Phi(r)\, \mathrm{d}r\Big) \cdot \Big(\int_{\partial\Omega_S} \underline{\bar{v}} \cdot \sigma_S(\alpha, \underline{u}) \cdot \mathbf{n}_S\, \mathrm{d}\sigma\Big).
\end{aligned}
$$

□

We may and will assume that $\Phi(r)$ does not vanish identically on $\mathbb{R}_+$ so that we can divide equation (4.14) by the integral $\int_0^\infty r^\alpha \Phi(r)\, \mathrm{d}r$ after the insertion of the expressions in Lemma 4.8. The following corollary states a special divergence theorem (Green's formula) on the unit sphere for vector functions $\underline{u}$ and $\underline{v}$ which are introduced by the relations in (4.15). It is a simple consequence of relation (4.14) and Lemma 4.8.

Corollary 4.9. *The relation*

$$
\begin{aligned}
&- \int_{\Omega_S} \underline{\bar{v}} \cdot \{\nabla_S \cdot \sigma_S(\alpha, \underline{u})\}\, \mathrm{d}\mathcal{S} - (\alpha-1) \int_{\Omega_S} \underline{\bar{v}} \cdot \{\underline{\mathbf{x}} \cdot \sigma_S(\alpha, \underline{u})\}\, \mathrm{d}\mathcal{S} \\
&= \int_{\Omega_S} \Big[\varepsilon_S(\underline{\bar{v}}) - (\alpha+1)\, \varepsilon_3(\underline{\bar{v}})\Big] : \sigma_S(\alpha, \underline{u})\, \mathrm{d}\mathcal{S} - \int_{\partial\Omega_S} \underline{\bar{v}} \cdot \sigma_S(\alpha, \underline{u}) \cdot \mathbf{n}_S\, \mathrm{d}\sigma \qquad (4.21)
\end{aligned}
$$

holds, where $\alpha \in \mathbb{C}$ and $\underline{u}, \underline{v} \in [\mathcal{H}^1(\Omega_S)]^3$ were introduced in (4.15).

We introduce the spaces

$$
H := [\mathcal{H}^0(\Omega_S)]^3, \qquad V := \{\underline{v} \in [\mathcal{H}^1(\Omega_S)]^3 \mid \underline{v} = 0 \text{ on } \Gamma_D\}
$$

of $\mathcal{H}^0$- and $\mathcal{H}^1$-vector functions. The presented theory includes the case, where the Dirichlet and Neumann boundary conditions are defined componentwise for the vector functions

$\underline{u} = \sum_{i=1}^{3} u_i \mathbf{e}_i$, so that $\underline{u} = 0$ on Γ_D is the compact notation for $u_i = 0$ on Γ_{D_i}, compare Remark 3.1. Due to Lemma 4.8, relation (4.14) yields

$$\int_{\Omega_S} \Big[\varepsilon_S(\bar{\underline{v}}) - (\alpha+1)\,\varepsilon_3(\bar{\underline{v}})\Big] : \sigma_S(\alpha, \underline{u})\,\mathrm{d}\mathcal{S} \;=\; 0 \tag{4.22}$$

for $\alpha \in \mathbb{C}$, $\underline{u}, \underline{v} \in V$. This is a *quadratic operator eigenvalue problem* for α and $\underline{u}$ which is defined on the unit sphere: We search for $\alpha \in \mathbb{C}$, $\underline{u} \in V$ such that equation (4.22) is fulfilled for all $\underline{v} \in V$. The quadratic structure of this eigenvalue problem can be read from

$$\begin{aligned}
&\int_{\Omega_S} \Big[\varepsilon_S(\bar{\underline{v}}) - (\alpha+1)\,\varepsilon_3(\bar{\underline{v}})\Big] : \sigma_S(\alpha, \underline{u})\,\mathrm{d}\mathcal{S} \\
&\quad= \int_{\Omega_S} \Big[\varepsilon_S(\bar{\underline{v}}) - (\alpha+1)\,\varepsilon_3(\bar{\underline{v}})\Big] : A : \Big[\varepsilon_S(\underline{u}) + \alpha\,\varepsilon_3(\underline{u})\Big]\,\mathrm{d}\mathcal{S} \\
&\quad= -\alpha(\alpha+1)\int_{\Omega_S} \varepsilon_3(\bar{\underline{v}}) : A : \varepsilon_3(\underline{u})\,\mathrm{d}\mathcal{S} - (\alpha+1)\int_{\Omega_S} \varepsilon_3(\bar{\underline{v}}) : A : \varepsilon_S(\underline{u})\,\mathrm{d}\mathcal{S} \\
&\qquad+ \alpha\int_{\Omega_S} \varepsilon_S(\bar{\underline{v}}) : A : \varepsilon_3(\underline{u})\,\mathrm{d}\mathcal{S} + \int_{\Omega_S} \varepsilon_S(\bar{\underline{v}}) : A : \varepsilon_S(\underline{u})\,\mathrm{d}\mathcal{S},
\end{aligned} \tag{4.23}$$

cf. [75]; there are certain sesquilinear forms $m_\alpha(\underline{u}, \underline{v})$, $g_\alpha(\underline{u}, \underline{v})$ and $k_\alpha(\underline{u}, \underline{v})$ so that

$$\int_{\Omega_S} \Big[\varepsilon_S(\bar{\underline{v}}) - (\alpha+1)\,\varepsilon_3(\bar{\underline{v}})\Big] : \sigma_S(\alpha, \underline{u})\,\mathrm{d}\mathcal{S} = \alpha^2\, m_\alpha(\underline{u}, \underline{v}) + \alpha\, g_\alpha(\underline{u}, \underline{v}) - k_\alpha(\underline{u}, \underline{v}).$$

We assume that the tensor A is bounded and elliptic. This means that there are positive constants M_1, M_2 such that

$$M_1 \int_{\Omega_S} (\tau^H : \tau)\,\mathrm{d}\mathcal{S} \le \int_{\Omega_S} \tau^H : A : \tau\,\mathrm{d}\mathcal{S} \le M_2 \int_{\Omega_S} (\tau^H : \tau)\,\mathrm{d}\mathcal{S} \tag{4.24}$$

for all (symmetric) second order tensor-valued functions τ. We will particularly replace τ by the symmetric spherical stress tensors ε_3 and ε_S, so that, in general, $\tau^\top = \tau = \bar{\tau}^H$. Moreover, we need that

$$\tau^\top : A : \sigma = \sigma^\top : A : \tau \tag{4.25}$$

for all (symmetric) second order tensors τ and σ.

We remark that the classical symmetry assumptions include more properties of A; their meaning will be discussed in Section 4.4.3. From the mathematical point of view, these additional symmetry assumptions on A are not necessary in the further analysis of the elasticity problem and do not at all follow from certain symmetry properties of the tensors ε or σ although usually claimed in the literature. When the classical symmetry properties are used, the restriction of the relation (4.24) to symmetric second order tensors $\tau = \tau^\top$ is essential, because A extinguishes all skew-symmetric tensors otherwise. The relations (4.24) and (4.25) imply that the form $(\tau, \sigma)_A = \int_{\Omega_S} \sigma^H : A : \tau\,\mathrm{d}\mathcal{S}$ defines an inner product on the space of (symmetric) second order tensor-valued functions.

The boundedness and ellipticity assumption (4.24) yields useful properties of the derived eigenvalue problem. In particular, the spectrum of problem (4.22) is symmetric with respect to the axes $\mathrm{Re}(\alpha) = -1/2$ and $\mathrm{Im}(\alpha) = 0$. This fact was mentioned, for instance, in [55, 64]. Its proof is not trivial; arguments from the theory of Fredholm operators have to be used [55, 88]. In Remark 4.15 (page 90), we will state a similar symmetry property for a related eigenvalue problem from which the spectral symmetry of problem (4.22) follows immediately.

For the numerical computation of the eigenvalues, it is useful to shift the spectrum so that it is symmetric with respect to the real and the imaginary axes. Then, the discretized eigenvalue problem is equivalent to a standard eigenvalue problem for a Hamiltonian or skew-Hamiltonian matrix, which can be solved efficiently with adapted Arnoldi or Lanczos algorithms [39, 15, 16, 72, 3, 106]. A Hamiltonian perturbation theory [52, 58, 103] reveals that the computation of the eigenvalues in this way becomes more stable than for non-structured methods. The desired shifting of the spectrum can be performed by substituting

$$\alpha = \lambda - \frac{1}{2}.$$

Then, a quadratic eigenvalue problem of the form

$$\lambda^2 m_\lambda(\underline{u}, \underline{v}) + \lambda g_\lambda(\underline{u}, \underline{v}) = k_\lambda(\underline{u}, \underline{v}) \tag{4.26}$$

is obtained for λ and $\underline{u}$ with appropriate Hermitian sesquilinear forms $m_\lambda(\underline{u}, \underline{v})$ and $k_\lambda(\underline{u}, \underline{v})$ and a skew-Hermitian sesquilinear form $g_\lambda(\underline{u}, \underline{v})$. Further properties of these sesquilinear forms are summarized in Theorem 4.11. The eigenvalue problems (4.22) and (4.26) are equivalent, this means that α is an eigenvalue of (4.22) if and only if $\lambda = \alpha + 1/2$ is an eigenvalue of (4.26). Moreover, $\underline{u} \in V$ is an eigenfunction of (4.22) corresponding to α if and only if $\underline{u}$ is an eigenfunction of (4.26) corresponding to $\lambda = \alpha + 1/2$.

The indices α and λ at the sesquilinear forms shall merely indicate that the formulation with α or λ is used, respectively; they do not represent specific numbers. Note that $m_\alpha(\underline{u}, \underline{v}) = m_\lambda(\underline{u}, \underline{v})$, whereas $g_\alpha(\underline{u}, \underline{v})$ and $g_\lambda(\underline{u}, \underline{v})$ as well as $k_\alpha(\underline{u}, \underline{v})$ and $k_\lambda(\underline{u}, \underline{v})$ differ significantly.

Definition 4.10. *In addition to the $\mathcal{H}^0$- and $\mathcal{H}^1$-norms*

$$\begin{aligned}
\|\!\lceil \underline{u} \rceil\!\|_{0,\Omega_S} &= \Big(\int_{\Omega_S} (\underline{\mathbf{x}\underline{u}})^H : (\underline{\mathbf{x}\underline{u}}) \,\mathrm{d}\mathcal{S}\Big)^{1/2} = \Big(\int_{\Omega_S} \bar{\underline{u}} \cdot \underline{u} \,\mathrm{d}\mathcal{S}\Big)^{1/2} \\
\textit{and} \quad \|\!\lceil \underline{u} \rceil\!\|_{1,\Omega_S} &= \Big(\frac{1}{4}\|\underline{u}\|_0^2 + |\underline{u}|_1^2\Big)^{1/2} \\
&= \Big(\int_{\Omega_S} (\nabla_S \underline{u} - \frac{1}{2}(\underline{\mathbf{x}\underline{u}}))^H : (\nabla_S \underline{u} - \frac{1}{2}(\underline{\mathbf{x}\underline{u}})) \,\mathrm{d}\mathcal{S}\Big)^{1/2} \\
\textit{with} \quad \lceil \underline{u} \rceil_{1,\Omega_S} &= \Big(\int_{\Omega_S} (\nabla_S \underline{u})^H : (\nabla_S \underline{u}) \,\mathrm{d}\mathcal{S}\Big)^{1/2}.
\end{aligned}$$

on H and V, we introduce the norms

$$\begin{aligned}
\|\underline{u}\|_{H_{\not\alpha}} &:= \Big(\int_{\Omega_S} \varepsilon_3(\bar{\underline{u}}) : \varepsilon_3(\underline{u}) \,\mathrm{d}\mathcal{S}\Big)^{1/2} \\
\textit{and} \quad \|\underline{u}\|_{V_{\not\alpha}} &:= \Big(\int_{\Omega_S} (\varepsilon_S(\bar{\underline{u}}) - \frac{1}{2}\varepsilon_3(\bar{\underline{u}})) : (\varepsilon_S(\underline{u}) - \frac{1}{2}\varepsilon_3(\underline{u})) \,\mathrm{d}\mathcal{S}\Big)^{1/2}
\end{aligned}$$

for $\underline{u}$ in H or V, respectively.

The $\mathcal{H}^0$- and $\mathcal{H}^1$-norms were defined in Section 2.4, see page 31. The validity of the representation in terms of $\underline{\mathbf{x}\underline{u}}$ and $\nabla_S \underline{u}$ can be verified with the use of tensor calculus. To this end, recall $\underline{\mathbf{x}} \cdot \underline{\mathbf{x}} = 1$ and note that $(\underline{\mathbf{x}}\underline{u})^H : (\nabla_S \underline{u}) = (\nabla_S \underline{u})^H : (\underline{\mathbf{x}}\underline{u}) = 0$ since $\mathbf{g}_S^i : \mathbf{g}^3 = 0$ for $i = 1, 2$. Therefore, the terms with the minus sign are extinguished in the second equality of the representation of the full $\mathcal{H}^1$-norm.

It is not clear at once that the expressions $\| \cdot \|_{H_{\not A}}$ and $\| \cdot \|_{V_{\not A}}$ define norms indeed. In particular, the positive definiteness is not obvious. This property becomes clear, when *(i)* of the Theorem 4.11 is proven. Like the standard norms $\|\lceil\cdot\rceil\|_{0,\Omega_S}$ and $\|\lceil\cdot\rceil\|_{1,\Omega_S}$, the norms $\| \cdot \|_{H_{\not A}}$ and $\| \cdot \|_{V_{\not A}}$ are independent of the material tensor A. To emphasize the independence of the material, they are indicated with the index $\not A$.

Theorem 4.11. *Let $\|\underline{u}\|_{H_{\not A}}$ and $\|\underline{u}\|_{V_{\not A}}$ be the norms introduced in Definition 4.10. The following assertions hold.*

(i) The norms $\|\underline{u}\|_{H_{\not A}}$ and $\|\underline{u}\|_{V_{\not A}}$ are equivalent to the norms $\|\lceil\underline{u}\rceil\|_0$ and $\|\lceil\underline{u}\rceil\|_1$, respectively. In particular,

$$\begin{aligned} \tfrac{1}{2}\|\lceil\underline{u}\rceil\|_0 &\leq \|\underline{u}\|_{H_{\not A}} \leq \|\lceil\underline{u}\rceil\|_0 \qquad \text{and} \\ c_K\|\lceil\underline{u}\rceil\|_1 &\leq \|\underline{u}\|_{V_{\not A}} \leq \|\lceil\underline{u}\rceil\|_1 \end{aligned}$$

with the constant c_K from Korn's inequality.

(ii) When the space V is equipped with the norm $\| \cdot \|_{V_{\not A}}$ and H is equipped with the norm $\| \cdot \|_{H_{\not A}}$, then V is compactly embedded in H.

(iii) The sesquilinear forms in (4.26) are given by

$$\begin{aligned} m_\lambda(\underline{u}, \underline{v}) &= \int_{\Omega_S} \varepsilon_3(\overline{\underline{v}}) : A : \varepsilon_3(\underline{u}) \, \mathrm{d}\mathcal{S} \\ k_\lambda(\underline{u}, \underline{v}) &= \int_{\Omega_S} \Big(\varepsilon_S(\overline{\underline{v}}) - \frac{1}{2}\varepsilon_3(\overline{\underline{v}})\Big) : A : \Big(\varepsilon_S(\underline{u}) - \frac{1}{2}\varepsilon_3(\underline{u})\Big) \, \mathrm{d}\mathcal{S} \\ g_\lambda(\underline{u}, \underline{v}) &= \overline{d(\underline{v}, \underline{u})} - d(\underline{u}, \underline{v}) \\ d(\underline{u}, \underline{v}) &= \int_{\Omega_S} \Big(\varepsilon_S(\overline{\underline{v}}) - \frac{1}{2}\varepsilon_3(\overline{\underline{v}})\Big) : A : \varepsilon_3(\underline{u}) \, \mathrm{d}\mathcal{S}, \end{aligned}$$

where $m_\lambda : H \times H \to \mathbb{C}$, $g_\lambda, k_\lambda : V \times V \to \mathbb{C}$ and $d : H \times V \to \mathbb{C}$. They satisfy

$$m_\lambda(\underline{u}, \underline{v}) = \overline{m_\lambda(\underline{v}, \underline{u})}, \quad g_\lambda(\underline{u}, \underline{v}) = -\overline{g_\lambda(\underline{v}, \underline{u})}, \quad k_\lambda(\underline{u}, \underline{v}) = \overline{k_\lambda(\underline{v}, \underline{u})}. \tag{4.27}$$

(iv) The sesquilinear forms defined in (iii) satisfy

$$\begin{aligned} M_1\|\underline{u}\|^2_{H_{\not A}} \leq m(\underline{u}, \underline{u}) &\leq M_2\|\underline{u}\|^2_{H_{\not A}} && \text{for } \underline{u} \in H, \\ |m(\underline{u}, \underline{v})| &\leq M_2\|\underline{u}\|_{H_{\not A}}\|\underline{v}\|_{H_{\not A}} && \text{for } \underline{u}, \underline{v} \in H, \\ M_1\|\underline{u}\|^2_{V_{\not A}} \leq k(\underline{u}, \underline{u}) &\leq M_2\|\underline{u}\|^2_{V_{\not A}} && \text{for } \underline{u} \in V, \\ |k(\underline{u}, \underline{v})| &\leq M_2\|\underline{u}\|_{V_{\not A}}\|\underline{v}\|_{V_{\not A}} && \text{for } \underline{u}, \underline{v} \in V, \\ |d(\underline{u}, \underline{v})| &\leq \sqrt{m(\underline{u}, \underline{u})}\sqrt{k(\underline{v}, \underline{v})} && \text{for } \underline{u} \in H,\ \underline{v} \in V, \\ |d(\underline{u}, \underline{v})| &\leq M_2\|\underline{u}\|_{H_{\not A}}\|\underline{v}\|_{V_{\not A}} && \text{for } \underline{u} \in H,\ \underline{v} \in V, \end{aligned}$$

where M_1 and M_2 are the constants in the boundedness and ellipticity assumption (4.24) of the material tensor A.

Before we prove this theorem, some remarks shall be made about the meaning of the properties stated in Theorem 4.11. Due to *(iv)*, the sesquilinear forms m_λ and k_λ define inner products on the spaces H and V, respectively. The norms $\|\cdot\|_{H_A}$ and $\|\cdot\|_{V_A}$ were introduced so that the constants in *(iv)* equal the smallest and the largest eigenvalues of the material tensor A. Appropriate norms are given by

$$\|\underline{u}\|_{H_A}^2 = m_\lambda(\underline{u},\underline{u}) \quad \text{for } \underline{u} \in H \quad \text{and} \quad \|\underline{u}\|_{V_A}^2 = k_\lambda(\underline{u},\underline{u}) \quad \text{for } \underline{u} \in V.$$

Since both of them are material-dependent, they are indicated with the index A.

Proof of Theorem 4.11. A detailed proof of the assertions of Theorem 4.11 is given in [75]. We will outline here only the main ideas.

(i) The assertion for the H-norm is easily seen from

$$\|\underline{u}\|_{H_A}^2 = \frac{1}{2}\int_{\Omega_S} \bar{\underline{u}}\cdot\underline{u} + |\underline{\mathbf{x}}\cdot\underline{u}|^2 \,\mathrm{d}S = \frac{1}{2}|\!\lceil \underline{u} \rceil\!|_0^2 + \frac{1}{2}\int_{\Omega_S} |\underline{\mathbf{x}}\cdot\underline{u}|^2 \,\mathrm{d}S,$$

and $0 \le |\underline{\mathbf{x}}\cdot\underline{u}|^2 \le (\underline{\mathbf{x}}\cdot\underline{\mathbf{x}})(\bar{\underline{u}}\cdot\underline{u}) = \bar{\underline{u}}\cdot\underline{u}$ (recall $\underline{\mathbf{x}}\cdot\underline{\mathbf{x}} = 1$). Hence, $\frac{1}{2}|\!\lceil \underline{u} \rceil\!|_0^2 \le \|\underline{u}\|_{H_A}^2 \le |\!\lceil \underline{u} \rceil\!|_0^2$.
Concerning the V-norm, we set $\gamma(\underline{u}) := \nabla_S \underline{u} - \frac{1}{2}\underline{\underline{\mathbf{x}u}}$ and obtain that

$$\begin{aligned}
\|\underline{u}\|_{V_A}^2 &= \frac{1}{4}\int_{\Omega_S} (\gamma(\bar{\underline{u}}) + \gamma(\bar{\underline{u}})^\top) : (\gamma(\underline{u}) + \gamma(\underline{u})^\top) \,\mathrm{d}S \\
&= \frac{1}{2}\int_{\Omega_S} \gamma(\bar{\underline{u}})^\top : \gamma(\underline{u}) + \gamma(\bar{\underline{u}}) : \gamma(\underline{u}) \,\mathrm{d}S \\
&= \frac{1}{2}\int_{\Omega_S} \gamma(\underline{u})^H : \gamma(\underline{u}) + \gamma(\bar{\underline{u}}) : \gamma(\underline{u}) \,\mathrm{d}S.
\end{aligned}$$

To estimate the second addend, we consider the inner product $(\tau,\sigma) := \int_{\Omega_S} \sigma^H : \tau \,\mathrm{d}S$; in particular, we have that $(\tau,\sigma) \le |(\tau,\sigma)| \le \sqrt{(\tau,\tau)}\sqrt{(\sigma,\sigma)}$ (Cauchy-Schwarz). We choose $\tau := \gamma(\underline{u})$ and $\sigma := \gamma(\underline{u})^\top$. Note that $(\gamma(\underline{u}),\gamma(\underline{u})^\top) = \int_{\Omega_S} \gamma(\bar{\underline{u}}) : \gamma(\underline{u}) \,\mathrm{d}S$ is always real. Simple calculation rules for tensors imply that

$$\begin{aligned}
\int_{\Omega_S} \gamma(\bar{\underline{u}}) : \gamma(\underline{u}) \,\mathrm{d}S &= (\gamma(\underline{u}),\gamma(\underline{u})^\top) \\
&\le \sqrt{(\gamma(\underline{u}),\gamma(\underline{u}))}\sqrt{(\gamma(\underline{u})^\top,\gamma(\underline{u})^\top)} \\
&= \Big(\int_{\Omega_S} \gamma(\underline{u})^H : \gamma(\underline{u}) \,\mathrm{d}S\Big)^{1/2}\Big(\int_{\Omega_S} (\gamma(\underline{u})^\top)^H : \gamma(\underline{u})^\top \,\mathrm{d}S\Big)^{1/2} \\
&= \Big(\int_{\Omega_S} \gamma(\underline{u})^H : \gamma(\underline{u}) \,\mathrm{d}S\Big)^{1/2}\Big(\int_{\Omega_S} \gamma(\underline{u})^H : \gamma(\underline{u}) \,\mathrm{d}S\Big)^{1/2} \\
&= \int_{\Omega_S} \gamma(\underline{u})^H : \gamma(\underline{u}) \,\mathrm{d}S.
\end{aligned}$$

Hence, $\|\underline{u}\|_{V_A}^2 \le \int_{\Omega_S} \gamma(\underline{u})^H : \gamma(\underline{u}) \,\mathrm{d}S = |\!\lceil \underline{u} \rceil\!|_1^2$.
To prove the estimate in the other direction, a modification of Korn's second inequality is needed which is valid also for domains without Dirichlet boundary. We refer to [75] for details and content ourselves with the proof of the positive definiteness of the

expressions $\varepsilon(\bar{\underline{U}}) : \varepsilon(\underline{U})$ and $(\varepsilon_S(\bar{\underline{u}}) - \frac{1}{2}\varepsilon_3(\bar{\underline{u}})) : (\varepsilon_S(\underline{u}) - \frac{1}{2}\varepsilon_3(\underline{u}))$, since these properties are not obvious. The main idea is to exploit the special structure of the function $\underline{u}$ resulting from the approach (1.3). Recall from (4.19) that $\underline{U} = r^{-1/2}\underline{u}$ satisfies $\varepsilon(\underline{U}) = r^{-3/2}(\varepsilon_S(\underline{u}) - \frac{1}{2}\varepsilon_3(\underline{u}))$, and therefore

$$\varepsilon(\bar{\underline{U}}) : \varepsilon(\underline{U}) \;=\; r^{-3}(\varepsilon_S(\bar{\underline{u}}) - \frac{1}{2}\varepsilon_3(\bar{\underline{u}})) : (\varepsilon_S(\underline{u}) - \frac{1}{2}\varepsilon_3(\underline{u})).$$

The right-hand side vanishes if and only if the strain tensor $\varepsilon(\underline{U})$ vanishes. Moreover, $\varepsilon(\underline{U}) = 0$ means that only rigid body motions are applied to $\underline{U}$, which results in $\underline{U}(\underline{\mathbf{X}}) = \underline{c}_0 + \underline{c}_1 \times \underline{\mathbf{X}}$ with $\underline{\mathbf{X}} = r\underline{\mathbf{x}}$ and constant vectors $\underline{c}_0$ and $\underline{c}_1$. Consequently,

$$r^{-1/2}\underline{u}(\mathbf{x}) \;=\; \underline{U}(\underline{\mathbf{X}}) \;=\; \underline{c}_0 + \underline{c}_1 \times r\underline{\mathbf{x}} \;=\; \underline{c}_0 + r(\underline{c}_1 \times \underline{\mathbf{x}}).$$

The comparison of the coefficients corresponding to the r-terms yields $\underline{u} \equiv 0$ and $\underline{U} \equiv 0$, which proves the desired positivity properties.

(ii) By Rellich's theorem, the space V equipped with the norm $|\!|\!\lceil\cdot\rceil\!|\!|_1$ is compactly embedded into the space H equipped with the norm $|\!|\!\lceil\cdot\rceil\!|\!|_0$. The assertion for the norms $|\!|\!\lceil\cdot\rceil\!|\!|_{V_\lambda}$ and $|\!|\!\lceil\cdot\rceil\!|\!|_{H_\lambda}$ follows from *(i)*.

(iii) The structures of the sesquilinear forms m_λ, g_λ and k_λ are obtained from direct computation using (4.23) and $\alpha = \lambda - 1/2$.

(iv) The ellipticity and boundedness properties of m_λ and k_λ follow from the ellipticity and boundedness properties of the material tensor A and from the Cauchy-Schwarz inequality. To prove the first estimate for d, we consider the form $(\tau,\sigma)_A = \int_{\Omega_S} \sigma^H : A : \tau \, \mathrm{d}S$ which defines an inner product on the space of symmetric second order tensor valued functions. The Cauchy-Schwarz inequality implies

$$|(\tau,\sigma)_A| \leq \sqrt{(\tau,\tau)_A}\sqrt{(\sigma,\sigma)_A}.$$

We choose $\tau := \varepsilon_3(\underline{u})$ and $\sigma := \varepsilon_S(\underline{v}) - \frac{1}{2}\varepsilon_3(\underline{v})$ and conclude that

$$|d(\underline{u},\underline{v})| = |(\tau,\sigma)_A| \leq \sqrt{m(\underline{u},\underline{u})}\sqrt{k(\underline{v},\underline{v})}.$$

The second estimate for d follows from the estimates for m_λ and k_λ.

□

Corollary 4.12. *The space V with the inner product $k_\lambda(\cdot,\cdot)$ and the corresponding norm $\|\underline{u}\|_{V_\lambda} = \sqrt{k_\lambda(\underline{u},\underline{u})}$ is a (complete) Hilbert space over $\mathbb{C}$.*

Likewise, the space H with the inner product $m_\lambda(\cdot,\cdot)$ is a (complete) Hilbert space over $\mathbb{C}$ with respect to the norm $\|\underline{u}\|_{H_\lambda} = \sqrt{m_\lambda(\underline{u},\underline{u})}$.

Remark 4.13. *The space V is the complexification of a real Hilbert space V_R which consists of the real-valued $\mathcal{H}^1$-functions. The sesquilinear forms m_λ, g_λ, k_λ are real-valued with respect to real arguments, because the material tensor A has real components. This means that $m_\lambda(\underline{u}_R,\underline{v}_R) \in \mathbb{R}$, $g_\lambda(\underline{u}_R,\underline{v}_R) \in \mathbb{R}$, $k_\lambda(\underline{u}_R,\underline{v}_R) \in \mathbb{R}$ for all $\underline{u}_R,\underline{v}_R \in V_R$. One easily checks that this is equivalent to*

$$\overline{m_\lambda(\underline{u},\underline{v})} = m_\lambda(\bar{\underline{u}},\bar{\underline{v}}), \qquad \overline{g_\lambda(\underline{u},\underline{v})} = g_\lambda(\bar{\underline{u}},\bar{\underline{v}}), \qquad \overline{k_\lambda(\underline{u},\underline{v})} = k_\lambda(\bar{\underline{u}},\bar{\underline{v}}) \tag{4.28}$$

for all $\underline{u},\underline{v} \in V$.

Lemma 4.14 ([55, Corollary 1.1.1]). *The spectrum of problem (4.26) consists of isolated eigenvalues with finite algebraic multiplicities and the only possible accumulation point at infinity.*

The assumptions of [55, Corollary 1.1.1] are satisfied due to the properties of the eigenvalue problem (4.26) which are summarized in Theorem 4.11, see also [88]. The original version of Lemma 4.14 with detailed proofs can be found in the work by Riesz [93, Satz 1, Satz 12].

Let $(\cdot,\cdot)_V : V \times V \to \mathbb{C}$ be an inner product on the space V. We define the operators $\mathcal{M}, \mathcal{G}, \mathcal{K} : V \to V$ by

$$\begin{array}{llcll} \mathcal{M}u \in V, & (\mathcal{M}u, v)_V & = & m_\lambda(u,v) & \forall v \in V, \\ \mathcal{G}u \in V, & (\mathcal{G}u, v)_V & = & g_\lambda(u,v) & \forall v \in V, \\ \mathcal{K}u \in V, & (\mathcal{K}u, v)_V & = & k_\lambda(u,v) & \forall v \in V. \end{array} \tag{4.29}$$

According to Lemma 4.1 and Theorem 4.11, these operators are well defined. The operator pencil $\mathcal{B}(\cdot) : \mathbb{C} \to \mathcal{L}(V,V)$,

$$\mathcal{B}(\lambda) := \mathcal{K} - \lambda\mathcal{G} - \lambda^2\mathcal{M}, \tag{4.30}$$

satisfies $(\mathcal{B}(\lambda)\underline{u}, \underline{v})_V = k_\lambda(\underline{u},\underline{v}) - \lambda g_\lambda(\underline{u},\underline{v}) - \lambda^2 m_\lambda(\underline{u},\underline{v})$. The adjoint operator pencil $\mathcal{B}^\star(\lambda) := [\mathcal{B}(\lambda)]^\star$ is given by

$$\mathcal{B}^\star(\lambda) = \mathcal{K}^\star - \bar{\lambda}\mathcal{G}^\star - \bar{\lambda}^2\mathcal{M}^\star.$$

Note that $\mathcal{M}^\star = \mathcal{M}$, $\mathcal{G}^\star = -\mathcal{G}$ and $\mathcal{K}^\star = \mathcal{K}$ due to (4.27). Consequently,

$$\mathcal{B}^\star(\lambda) = \mathcal{B}(-\bar{\lambda}). \tag{4.31}$$

It follows from Theorem 4.11 that the operators $\mathcal{M}$ and $\mathcal{G}$ are compact, see also [88, Section 3.3]. By Lemma 2.3, the operator $\mathcal{B}(\lambda)$ is a Fredholm operator with the index zero for each $\lambda \in \mathbb{C}$ if and only if $\mathcal{K}$ is invertible. In the following, we will choose

$$(\underline{u}, \underline{v})_V = k_\lambda(\underline{u}, \underline{v}) \quad \forall \underline{u}, \underline{v} \in V.$$

Then, $\mathcal{K}$ is the identity operator and therefore invertible and consequently, $\mathcal{B}(\lambda)$ is a Fredholm operator for each $\lambda \in \mathbb{C}$.

An important fact is here that $k_\lambda(\cdot,\cdot)$ defines an inner product, in particular, that the form $k_\lambda(\cdot,\cdot)$ is positive definite. Note that the form $k_\alpha(\cdot,\cdot)$ in (4.22) is not positive definite. This is why we consider the operator pencil and its derivatives with respect to λ.

Remark 4.15. *The eigenvalues of problem (4.26) appear in quadruplets $(\lambda,\ \bar{\lambda},\ -\lambda,\ -\bar{\lambda})$ or in real or purely imaginary pairs $(\lambda, -\lambda)$. This property is called* Hamiltonian eigenvalue symmetry *motivated by the spectral properties of a Hamiltonian matrix (cf. [73]). The appearance of the pairs $(\lambda,\ \bar{\lambda})$ is easily seen from (4.28) and*

$$\lambda^2\, m_\lambda(\underline{u},\underline{v}) + \lambda\, g_\lambda(\underline{u},\underline{v}) - k_\lambda(\underline{u},\underline{v}) = 0 \quad \forall \underline{v} \in V$$

$$\Longleftrightarrow$$

$$\bar{\lambda}^2\, m_\lambda(\bar{\underline{u}},\bar{\underline{v}}) + \bar{\lambda}\, g_\lambda(\bar{\underline{u}},\bar{\underline{v}}) - k_\lambda(\bar{\underline{u}},\bar{\underline{v}}) = 0 \quad \forall \underline{v} \in V$$

$$\Longleftrightarrow$$

$$\bar{\lambda}^2\, m_\lambda(\bar{\underline{u}},\underline{v}) + \bar{\lambda}\, g_\lambda(\bar{\underline{u}},\underline{v}) - k_\lambda(\bar{\underline{u}},\underline{v}) = 0 \quad \forall \underline{v} \in V.$$

The eigenfunction corresponding to $\bar{\lambda}$ is the conjugate complex of the eigenfunction corresponding to λ. Furthermore, we obtain from (4.31) that λ is an eigenvalue of (4.26) if and only if $-\bar{\lambda}$ is an eigenvalue of (4.26), cf. Remark 2.7, see also [88, 55].

Singularity exponents are always denoted by α (eigenvalues of problem (4.22)), whereas λ denotes the value $\alpha + 1/2$ or, synonymously, an eigenvalue of the shifted eigenvalue problem (4.26). One readily verifies that $\lambda = 0$ is not an eigenvalue of problem (4.26), since $k_\lambda(\cdot,\cdot)$ is positive definite. This means that $\alpha = -1/2$ is not an eigenvalue of problem (4.22), whereas $\alpha = 0$ might be. For the Lamé system of isotropic elasticity, it was shown in [55, Theorems 3.2.1, 4.1.1, 4.3.1] that the line $\mathrm{Re}(\alpha) = -1/2$ does not contain eigenvalues. According to [55, Theorem 4.4.1], the same is true for the Neumann problem of an angular crack in an anisotropic elastic space. From the way the proof is formulated, it can be presumed that there are generally no eigenvalues on the axis $\mathrm{Re}(\alpha) = -1/2$. It would go beyond the scope of this thesis to prove this for general spherical domains and general boundary conditions.

Concerning condition (2.17) in Theorem 2.8, let λ_0 be an eigenvalue of problem (4.26) with the associated eigenfunction $\underline{u}_0$. Then, we have to show that

$$(\mathcal{B}'(\lambda_0)\underline{u}_0, \underline{u}_0)_V = -g_\lambda(\underline{u}_0, \underline{u}_0) - 2\lambda_0 m_\lambda(\underline{u}_0, \underline{u}_0) \neq 0.$$

Due to Theorem 4.11, we have $m_\lambda(\underline{u}_0, \underline{u}_0) > 0$ and $g_\lambda(\underline{u}_0, \underline{u}_0) = it$ for some $t \in \mathbb{R}$. Thus, $(\mathcal{B}'(\lambda_0)\underline{u}_0, \underline{u}_0)_V$ vanishes if and only if λ_0 is purely imaginary. To ensure that $\mathcal{B}(\cdot)$ satisfies the condition (2.17), we have to exclude eigenvalues on the line $\mathrm{Re}(\lambda) = 0$ (or $\mathrm{Re}(\alpha) = -1/2$). For this reason, we restrict our considerations to eigenvalues of problem (4.26) which lie away from the imaginary axis.

Let $(\cdot,\cdot)_H : V \times V \to \mathbb{C}$ be a second inner product on V. Under the assumption $(\underline{u}, \bar{\underline{u}})_H = 1$, problem (4.26) is equivalent to $F(\lambda, \underline{u}) = 0$, where $F : \mathbb{C} \times V \to \mathbb{C} \times V^\star$ is given by

$$\langle F([\lambda, \underline{u}]), [\bar{\mu}, \bar{\underline{v}}]\rangle = k_\lambda(\underline{u}, \underline{v}) - \lambda g_\lambda(\underline{u}, \underline{v}) - \lambda^2 m_\lambda(\underline{u}, \underline{v}) + \bar{\mu}((\underline{u}, \bar{\underline{u}})_H - 1)$$

for $\lambda, \mu \in \mathbb{C}$, $\underline{u}, \underline{v} \in V$. Since (4.22) and (4.26) are equivalent, F can be written in the form

$$\begin{aligned}\langle F([\lambda, \underline{u}]), [\bar{\mu}, \bar{\underline{v}}]\rangle &= \langle F([\alpha + 1/2, \underline{u}]), [\bar{\mu}, \bar{\underline{v}}]\rangle \\ &= \int_{\Omega_S} [\varepsilon_S(\bar{\underline{v}}) - (\alpha+1)\varepsilon_3(\bar{\underline{v}})] : \sigma_S(\alpha, \underline{u})\,\mathrm{d}S + \bar{\mu}((\underline{u}, \bar{\underline{u}})_H - 1). \quad (4.32)\end{aligned}$$

Remark 4.16. *The representation of F with respect to α is needed (or more convenient) to compute the element and edge residuals.*

The specific choice of the inner product $(\cdot,\cdot)_H : V \times V \to \mathbb{C}$ is irrelevant. In accordance with the notation of Section 2.3, the index H is used to indicate that the space V is not necessarily complete with respect to the norm induced by this inner product. In general, there is no connection to the space H. Usually, however, one will choose an inner product which is defined on $H \times H$ like $(\underline{u}, \underline{v})_H = m_\lambda(\underline{u}, \underline{v})$ or the material-independent inner product $(\underline{u}, \underline{v})_H = \int_{\Omega_S} \varepsilon_3(\bar{\underline{v}}) : \varepsilon_3(\underline{u})\,\mathrm{d}S$. Alternatively, the V-inner products $(\underline{u}, \underline{v})_H = k_\lambda(\underline{u}, \underline{v})$ or $(\underline{u}, \underline{v})_H = \int_{\Omega_S} (\varepsilon_S(\bar{\underline{v}}) - \frac{1}{2}\varepsilon_3(\bar{\underline{v}})) : (\varepsilon_S(\underline{u}) - \frac{1}{2}\varepsilon_3(\underline{u}))\,\mathrm{d}S$ can be used. All we need is $(\underline{u}, \underline{u})_H \leq C_{H,V}^2(\underline{u}, \underline{u})_V = C_{H,V}^2 k_\lambda(\underline{u}, \underline{u})$ for all $\underline{u} \in V$ and some constant $C_{H,V} > 0$. Remarks regarding the implementation are given in Section 4.4.2.

According to Remark 2.9, the condition $(u, \bar{\underline{u}})_H = 1$ has an influence only on the constants in the upper error bounds. For the lower error bound in Theorem 3.23, it is not necessary to assume that $(u, \bar{\underline{u}})_V \neq 0$.

Possible eigenfunctions satisfying $(\underline{u}, \bar{\underline{u}})_H = 0$ are skipped. An interpretation of this exception regarding the specific application (linear elasticity) has not been found yet. Note that $m_\lambda(\underline{u}_0, \bar{\underline{u}}_0) = 0$ if and only if $k_\lambda(\underline{u}_0, \bar{\underline{u}}_0) = 0$, if $\underline{u}_0$ is an eigenfunction of problem (4.26). This is easily seen from $g_\lambda(\underline{u}_0, \bar{\underline{u}}_0) = 0$ according to relation (4.27) and from

$k_\lambda(\underline{u}_0, \bar{\underline{u}}_0) = \lambda_0^2 m_\lambda(\underline{u}_0, \bar{\underline{u}}_0) + \lambda_0 g_\lambda(\underline{u}_0, \bar{\underline{u}}_0) = \lambda_0^2 m_\lambda(\underline{u}_0, \bar{\underline{u}}_0)$ *according to (4.26), where* $\lambda_0 \neq 0$ *is the eigenvalue corresponding to* $\underline{u}_0$.

Now let $\mathcal{T}_h$ be a triangulation of Ω_S as defined in Sections 3.1 and 3.2, and let V_h be an appropriate finite element space, see page 37. The modified divergence theorem (4.21) still holds, when we replace Ω_S by an arbitrary connected subdomain $T \subset \Omega_S$. With (4.32) and (4.1), we conclude for each $[\lambda_h, \underline{u}_h] \in \mathbb{C} \times V_h$ with $(\underline{u}_h, \bar{\underline{u}}_h)_H = 1$. that

$$\begin{aligned}
\langle F([\lambda_h, \underline{u}_h]), [\bar{\mu}, \bar{\underline{v}}]\rangle &= \langle F([\alpha_h + 1/2, \underline{u}_h]), [\bar{\mu}, \bar{\underline{v}}]\rangle \\
&= \sum_{T\in\mathcal{T}_h} \Big[-\int_T \bar{\underline{v}} \cdot \{\nabla_S \cdot \sigma_S(\alpha_h, \underline{u}_h)\}\, \mathrm{d}\mathcal{S} - (\alpha_h - 1) \int_T \bar{\underline{v}} \cdot \{\underline{\mathbf{x}} \cdot \sigma_S(\alpha_h, \underline{u}_h)\}\, \mathrm{d}\mathcal{S} \\
&\qquad + \sum_{E\in\mathcal{E}(T)} \int_E \bar{\underline{v}} \cdot \sigma_S(\alpha_h, \underline{u}_h) \cdot \mathbf{n}_{T,E}\, \mathrm{d}\sigma \Big] \\
&= \sum_{T\in\mathcal{T}_h} \Big[-\int_T \bar{\underline{v}} \cdot \{\nabla_S \cdot \sigma_S(\alpha_h, \underline{u}_h)\}\, \mathrm{d}\mathcal{S} - (\alpha_h - 1) \int_T \bar{\underline{v}} \cdot \{\underline{\mathbf{x}} \cdot \sigma_S(\alpha_h, \underline{u}_h)\}\, \mathrm{d}\mathcal{S} \Big] \\
&\quad - \sum_{E\in\mathcal{E}_{h,\Omega_S}} \int_E \bar{\underline{v}} \cdot \lfloor \sigma_S(\alpha_h, \underline{u}_h) \cdot \mathbf{n}_E \rfloor_E\, \mathrm{d}\sigma + \sum_{E\in\mathcal{E}_{h,N}} \int_E \bar{\underline{v}} \cdot \sigma_S(\alpha_h, \underline{u}_h) \cdot \mathbf{n}_E\, \mathrm{d}\sigma
\end{aligned} \tag{4.33}$$

for all $[\mu, \underline{v}] \in \mathbb{K} \times V$. The element and edge residuals in (3.12) are therefore given by

$$\begin{aligned}
\underline{R}_T &:= \{-\nabla_S \cdot \sigma_S(\alpha_h, \underline{u}_h) - (\alpha_h - 1)\, \underline{\mathbf{x}} \cdot \sigma_S(\alpha_h, \underline{u}_h)\}|_T \\
&= \{-\nabla_S \cdot \sigma_S(\lambda_h - 1/2, \underline{u}_h) - (\lambda_h - 3/2)\, \underline{\mathbf{x}} \cdot \sigma_S(\lambda_h - 1/2, \underline{u}_h)\}|_T, \qquad (4.34)\\
\underline{R}_E &:= \lfloor \mathbf{n}_E \cdot \sigma_S(\alpha_h, \underline{u}_h) \rfloor_E = \lfloor \mathbf{n}_E \cdot \sigma_S(\lambda_h - 1/2, \underline{u}_h) \rfloor_E, \\
\underline{R}_N &:= \mathbf{n}_E \cdot \sigma_S(\alpha_h, \underline{u}_h) = \mathbf{n}_E \cdot \sigma_S(\lambda_h - 1/2, \underline{u}_h),
\end{aligned}$$

for $[\alpha_h, \underline{u}_h] = [\lambda_h - 1/2, \underline{u}_h] \in \mathbb{C} \times V_h$.

To obtain the residuals for the adjoint eigenvalue problem, we recall $\mathcal{B}(\lambda) = \mathcal{B}^\star(-\bar{\lambda})$ for all $\lambda \in \mathbb{C}$, see (4.31). When λ is an eigenvalue of the adjoint pencil $\mathcal{B}^\star(\cdot)$ with the eigenfunction $\underline{u}^\star$, then $-\bar{\lambda}$ is an eigenvalue of $\mathcal{B}(\cdot)$ with the same eigenfunction. This means that we simply have to compute an eigenfunction $\underline{u}^\star$ corresponding to $-\bar{\lambda}$ with $(\underline{u}^\star, \bar{\underline{u}}^\star)_H = 1$ instead of solving the adjoint eigenvalue problem. The residuals for the adjoint problem read

$$\begin{aligned}
\underline{R}_T^\star &:= \{-\nabla_S \cdot \sigma_S(-\bar{\lambda}_h - 1/2, \underline{u}_h^\star) - (-\bar{\lambda}_h - 3/2)\, \underline{\mathbf{x}} \cdot \sigma_S(-\bar{\lambda}_h - 1/2, \underline{u}_h^\star)\}|_T \\
&= \{-\nabla_S \cdot \sigma_S(-1 - \bar{\alpha}_h, \underline{u}_h^\star) - (-\bar{\alpha}_h - 2)\, \underline{\mathbf{x}} \cdot \sigma_S(-1 - \bar{\alpha}_h, \underline{u}_h^\star)\}|_T, \\
\underline{R}_E^\star &:= \lfloor \mathbf{n}_E \cdot \sigma_S(-\bar{\lambda}_h - 1/2, \underline{u}_h^\star) \rfloor_E = \lfloor \mathbf{n}_E \cdot \sigma_S(-1 - \bar{\alpha}_h, \underline{u}_h^\star) \rfloor_E, \\
\underline{R}_N &:= \mathbf{n}_E \cdot \sigma_S(-\bar{\lambda}_h - 1/2, \underline{u}_h^\star) = \mathbf{n}_E \cdot \sigma_S(-1 - \bar{\alpha}_h, \underline{u}_h^\star),
\end{aligned}$$

The error estimators $\tilde{\eta}_T$, η_T and the approximation errors ε_T are defined in (3.13), (3.15) and (3.16). The appropriate errors $\tilde{\eta}_T^\star$, $\eta_T^\star$ and $\varepsilon_T^\star$ for the adjoint problem are defined likewise with $\underline{R}_T$, $\underline{R}_E$ and $\underline{R}_N$ replaced by $\underline{R}_T^\star$, $\underline{R}_E^\star$ and $\underline{R}_N^\star$, respectively. We summarize the results of Sections 3.4, 3.5 and 3.6 in the following Theorem.

Theorem 4.17. *Let* $\lambda_0 \in \mathbb{C}$ *with* $\mathrm{Re}(\lambda) \neq 0$ *be a simple eigenvalue of problem (4.26), and let* $\underline{u}_0 \in V$ *be a corresponding eigenfunction with* $(\underline{u}_0, \bar{\underline{u}}_0)_H = 1$. *Let* $\underline{u}_0^\star \in V$ *be an eigenfunction of (4.26) corresponding to* $-\bar{\lambda}_0$ *with* $(\underline{u}_0^\star, \bar{\underline{u}}_0^\star)_H = 1$. *Furthermore, let* $[\lambda_h, \underline{u}_h] \in \mathbb{C} \times V_h$

and $[-\bar{\lambda}_h, \underline{u}_h^\star] \in \mathbb{C} \times V_h$ *with* $(\underline{u}_h, \bar{\underline{u}}_h)_H = 1$ *and* $(\underline{u}_h^\star, \bar{\underline{u}}_h^\star)_H = 1$ *be two eigenpairs of the discretized eigenvalue problem (3.4) which are close enough to* $[\lambda_0, \underline{u}_0]$ *and* $[-\bar{\lambda}_0, \underline{u}_0^\star]$ *so that the assumptions of Lemma 2.4 are satisfied. Then the estimates*

$$|\lambda_0 - \lambda_h| + \llbracket \underline{u}_0 - \underline{u}_h \rrbracket_{1,\Omega_S} \lesssim \Big\{\sum_{T\in\mathcal{T}_h} \tilde{\eta}_T^2\Big\}^{1/2} \leq \Big\{\sum_{T\in\mathcal{T}_h} \eta_T^2\Big\}^{1/2} + \Big\{\sum_{T\in\mathcal{T}_h} \varepsilon_T^2\Big\}^{1/2}$$

$$\Big\{\sum_{T\in\mathcal{T}_h} \eta_T^2\Big\}^{1/2} \lesssim |\lambda_0 - \lambda_h| + \llbracket \underline{u}_0 - \underline{u}_h \rrbracket_{1,\Omega_S} + \Big\{\sum_{T\in\mathcal{T}_h} \varepsilon_T^2\Big\}^{1/2},$$

$$\begin{aligned} |\lambda_0 - \lambda_h| &\lesssim \Big\{\sum_{T\in\mathcal{T}_h} \tilde{\eta}_T^2\Big\}^{1/2} \Big\{\sum_{T\in\mathcal{T}_h} \tilde{\eta}_T^{\star 2}\Big\}^{1/2} \\ &\leq \Big(\Big\{\sum_{T\in\mathcal{T}_h} \eta_T^2 + \varepsilon_T^2\Big\}^{1/2} \Big(\Big\{\sum_{T\in\mathcal{T}_h} \eta_T^{\star 2} + \varepsilon_T^{\star 2}\Big\}^{1/2} \\ &\leq \Big(\Big\{\sum_{T\in\mathcal{T}_h} \eta_T^2\Big\}^{1/2} + \Big\{\sum_{T\in\mathcal{T}_h} \varepsilon_T^2\Big\}^{1/2}\Big) \Big(\Big\{\sum_{T\in\mathcal{T}_h} \eta_T^{\star 2}\Big\}^{1/2} + \Big\{\sum_{T\in\mathcal{T}_h} \varepsilon_T^{\star 2}\Big\}^{1/2}\Big) \end{aligned}$$

hold.

The constants which are involved in these estimates are summarized in Section 3.7. Their dependence on the material parameters (that is on the tensor A) will be discussed in Section 4.3.2.

A priori error estimates [7] for the eigenvalue problem (4.26) show that we can expect quadratic convergence order for the eigenvalues and linear convergence order for the eigenfunctions when graded meshes are used near the corners of the domain. In particular,

$$\begin{aligned} |\lambda_0 - \lambda_h| &\lesssim h^{2/\kappa} |\lambda_0|^{4/\kappa}, \\ \llbracket \underline{u}_0 - \underline{u}_h \rrbracket_{1,\Omega_S} &\lesssim h^{\gamma} |\lambda_0|^{2/\gamma}, \quad \gamma = \min\{1, 2/\kappa\}, \end{aligned}$$

where κ is length of the longest Jordan chain corresponding to λ_0, in our case $\kappa = 1$, and where h is the global mesh size which relates to the number N of degrees of freedom by $N \sim h^{-2}$, see [7, Corollary 4.16]. As the number λ_0 does not depend on h, it can be considered as a constant factor. Large eigenvalues λ_0, however, diminish the accuracy of this approximation. The results of Theorem 4.17 are confirmed by these two estimates.

When the mesh is refined adaptively on the basis of the computed error estimator, it is not necessary to start with graded meshes, because the error is largest near the angular points in the domain Ω_S, so that the mesh is particularly refined there anyway, see Section 4.5.

4.3.2 Constants: Dependence on the material parameters

In Section 3.7, we saw that the constants in the a posteriori error estimates (see Theorem 4.17) depend on fixed constants stemming from the isotropy condition on the mesh (see Section 3.2) and from auxiliary results (spherical trace theorem, Poincaré-type inequality, estimates for the bubble functions etc.) as well as on the terms $\|DF([\lambda_0, \underline{u}_0])^{-1}\|_{\mathcal{L}(\mathbb{C}\times V^\star, \mathbb{C}\times V)}$ and $\|DF([\lambda_0, \underline{u}_0])\|_{\mathcal{L}(\mathbb{C}\times V, \mathbb{C}\times V^\star)}$. Estimates of the last two terms are given in Lemma 2.12. This section is aimed at deriving more precise dependencies of these two constants on λ_0 and $\underline{u}_0$ and on the material parameters given by the tensor A.

Recall $\|\underline{u}\|_{V_A}^2 = k_\lambda(\underline{u},\underline{u})$ and $\|\underline{u}\|_{H_A}^2 = m_\lambda(\underline{u},\underline{u})$, whereas $\|\underline{u}\|_{V_{\not A}}$ and $\|\underline{u}\|_{H_{\not A}}$ are the material-independent norms which were defined in Definition 4.10. Moreover, let $(\cdot,\cdot)_H$ be a second inner product, for instance, $(\underline{u},\underline{v})_H = m_\lambda(\underline{u},\underline{v})$. We abbreviate $\|\underline{u}\|_H := \sqrt{(\underline{u},\underline{u})_H}$. According to Theorem 4.11, the inequalities

$$\sqrt{M_1}\|\underline{u}\|_{V_{\not A}} \le \|\underline{u}\|_{V_A} \le \sqrt{M_2}\|\underline{u}\|_{V_{\not A}}, \qquad \sqrt{M_1}\|\underline{u}\|_{H_{\not A}} \le \|\underline{u}\|_{H_A} \le \sqrt{M_2}\|\underline{u}\|_{H_{\not A}},$$

$$\|\underline{u}\|_{H_{\not A}} \le \|\lceil \underline{u}\rceil\|_{0,\Omega_S} \le 2\|\lceil \underline{u}\rceil\|_{1,\Omega_S} \le \frac{2}{c_K}\|\underline{u}\|_{V_{\not A}},$$

hold. Let $C_{H,V}$ and C_{H_A,V_A} be the constants in $\|\underline{u}\|_H \le C_{H,V}\|\underline{u}\|_V = C_{H,V}\|\underline{u}\|_{V_A}$ and $\|\underline{u}\|_{H_A} \le C_{H_A,V_A}\|\underline{u}\|_{V_A}$. If $(\underline{u},\underline{u})_H = m_\lambda(\underline{u},\underline{v})$, then $C_{H,V} = C_{H_A,V_A}$. Moreover, $C_{H_A,V_A} \le 2/c_K \cdot \sqrt{M_2/M_1}$.

In Lemma 2.12, we learnt that

$$\|\mathrm{D}F([\lambda_0,\underline{u}_0])\|_{\mathcal{L}(\mathbb{C}\times V,\mathbb{C}\times V^\star)} \le \|\mathcal{B}'(\lambda_0)\,\underline{u}_0\|_{V_A} + \|\mathcal{B}(\lambda_0)\|_{\mathcal{L}(V,V)} + 2\,\|\underline{u}_0\|_H$$

and

$$\|\mathrm{D}F([\lambda_0,u_0])^{-1}([r,f])\|_{\mathbb{C}\times V} \le \|\underline{u}_0^\star\|_{V_A}|k_\lambda(\mathcal{B}'(\lambda_0)\underline{u}_0,\underline{u}_0^\star)^{-1}| + \frac{1}{2}\|\underline{u}_0\|_{V_A} + \sup_{\substack{f\in V^\star\\ \|f\|_{V^\star}\le 1}} \|\underline{u}_{w_\rho}\|_{V_A}.$$

According to (4.30), the operator pencil $\mathcal{B}(\cdot):\mathbb{C}\to\mathcal{L}(V,V)$ is given by $\mathcal{B}(\lambda) = \mathcal{K} - \lambda\mathcal{G} - \lambda^2\mathcal{M}$ and satisfies $\mathcal{B}'(\lambda) = -\mathcal{G} - 2\lambda\mathcal{M}$, where the operators $\mathcal{M}$, $\mathcal{G}$ and $\mathcal{K}$ are defined in (4.29). Recall that the inner product on V was chosen such that $\mathcal{K}$ is the identity operator in V. We conclude from Theorem 4.11 that

$$\begin{aligned}
\|\mathcal{M}\underline{u}\|_{V_A}^2 &= k_\lambda(\mathcal{M}\underline{u},\mathcal{M}\underline{u}) = m_\lambda(\underline{u},\mathcal{M}\underline{u}) \le M_2\|\underline{u}\|_{H_A}\|\mathcal{M}\underline{u}\|_{H_A}\\
&\le M_2 C_{H_A,V_A}\|\underline{u}\|_{H_A}\|\mathcal{M}\underline{u}\|_{V_A},\\
\|\mathcal{G}\underline{u}\|_{V_A}^2 &= k_\lambda(\mathcal{G}\underline{u},\mathcal{G}\underline{u}) = g(\underline{u},\mathcal{G}\underline{u}) \le M_2(\|\mathcal{G}\underline{u}\|_{H_A}\|\underline{u}\|_{V_A} + \|\mathcal{G}\underline{u}\|_{V_A}\|\underline{u}\|_{H_A})\\
&\le M_2(\|\underline{u}\|_{H_A} + C_{H_A,V_A}\|\underline{u}\|_{V_A})\|\mathcal{G}\underline{u}\|_{V_A},\\
\|\mathcal{K}\underline{u}\|_{V_A}^2 &= \|\underline{u}\|_{V_A},
\end{aligned}$$

and thus, $\|\mathcal{M}\underline{u}\|_{V_A} \le M_2 C_{H_A,V_A}\|\underline{u}\|_{H_A}$ and $\|\mathcal{G}\underline{u}\|_{V_A} \le M_2(\|\underline{u}\|_{H_A} + C_{H_A,V_A}\|\underline{u}\|_{V_A})$. Furthermore,

$$\mathcal{B}'(\lambda_0)\underline{u}_0 = \frac{1}{\lambda_0}(-\mathcal{K} + \mathcal{K} - \lambda_0\mathcal{G} - 2\lambda_0^2\mathcal{M})\underline{u}_0 = -\frac{1}{\lambda_0}(\mathcal{K} + \lambda_0^2\mathcal{M})\underline{u}_0,$$

since $[\lambda_0,\underline{u}_0]$ is an eigenpair of problem (2.13). Hence,

$$\begin{aligned}
\|\mathcal{B}'(\lambda_0)\underline{u}_0\|_{V_A} &\le \frac{1}{|\lambda_0|}\|\underline{u}_0\|_{V_A} + |\lambda_0|\,\|\mathcal{M}\underline{u}_0\|_{V_A} \le \frac{1}{|\lambda_0|}\|\underline{u}_0\|_{V_A} + M_2 C_{H_A,V_A}\,|\lambda_0|\,\|\underline{u}_0\|_{H_A}\\
&\le \frac{M_2}{|\lambda_0|}\|\underline{u}_0\|_{V_{\not A}} + M_2^2 C_{H_A,V_A}\,|\lambda_0|\,\|\underline{u}_0\|_{H_{\not A}},\\
\|\mathcal{B}(\lambda_0)\|_{\mathcal{L}(V,V)} &= \sup_{\substack{\underline{u}\in V\\ \|\underline{u}\|_{V_A}=1}} \|\mathcal{B}(\lambda_0)\underline{u}\|_{V_A} \le \sup_{\substack{\underline{u}\in V\\ \|\underline{u}\|_{V_A}=1}} \|\underline{u}\|_V + |\lambda_0|\,\|\mathcal{G}\underline{u}\|_{V_A} + |\lambda_0|^2\|\mathcal{M}\underline{u}\|_{V_A}\\
&\le 1 + M_2|\lambda_0|(C_{H_A,V_A}^2 + 1) + M_2|\lambda_0|^2 C_{H_A,V_A}^2,
\end{aligned}$$

so that

$$\begin{aligned}\|DF([\lambda_0,\underline{u}_0])\|_{\mathcal{L}(\mathbb{C}\times V,\mathbb{C}\times V^\star)} \leq\ & \frac{M_2}{|\lambda_0|}\|\underline{u}_0\|_{V_A^\star} + M_2^2 C_{H_A,V_A}\,|\lambda_0|\,\|\underline{u}_0\|_{H_A^\star} \\ & + 1 + M_2|\lambda_0|(C_{H_A,V_A}^2+1) + M_2|\lambda_0|^2 C_{H_A,V_A}^2 + 2\|\underline{u}_0\|_H .\end{aligned}$$

The norms $\|\underline{u}_0\|_{V_A^\star}$ and $\|\underline{u}_0\|_{H_A^\star}$ are independent of the material. Moreover, $\|\underline{u}_0\|_H \leq \sqrt{M_2}C_{H,V}\|\underline{u}_0\|_{V_A^\star}$. Thus, the constant $\|DF([\lambda_0,\underline{u}_0])\|_{\mathcal{L}(\mathbb{C}\times V,\mathbb{C}\times V^\star)}$ in the lower error bound depends mainly on M_2 and on $|\lambda_0|$ and u_0. It can be presumed that the constant C_{H_A,V_A} depends in addition on $1/M_1$ and $1/c_K$.

For the constant $\|DF([\lambda_0,\underline{u}_0])^{-1}\|_{\mathcal{L}(\mathbb{C}\times V^\star,\mathbb{C}\times V)}$ in the upper error bound, a major dependence on $1/M_1$ can be expected. The explicit constant is not computable.

The constants M_1 and M_2 correspond to the minimum and maximum eigenvalues of the material tensor. For isotropic material with the Lamé constants $\hat{\mu}$ and $\hat{\lambda}$, the material tensor A is usually given in the form

$$\sigma = A:\varepsilon = \hat{\mu}\,(\varepsilon+\varepsilon^\top) + \hat{\lambda}\,\mathrm{tr}\,\varepsilon = 2\hat{\mu}\,\varepsilon(\underline{U}) + \hat{\lambda}\,\mathrm{tr}\,\varepsilon$$

and possesses the eigenvalue $2\hat{\mu}$ with the algebraic multiplicity 5 and the simple eigenvalue $2\hat{\mu}+3\hat{\lambda}$, see, for example, [81]. The Lamé constants are related to the Poisson ratio ν and Young's modulus E by

$$\hat{\lambda} = \frac{2\hat{\mu}\nu}{1-2\nu} = \frac{E\nu}{(1+\nu)(1-2\nu)}, \qquad \hat{\mu} = \frac{\hat{\lambda}(1-2\nu)}{2\nu} = \frac{E}{2(1+\nu)},$$

compare [81, Table 3.1.1]. We remark that in the formulation (4.22) of the eigenvalue problem, one of these constants can be eliminated, for instance, by multiplying the equation with $(1+\nu)(1-2\nu)/E$. Then, $M_1 = 1-2\nu$, $M_2 = 1+\nu$, where $\nu \in (0, 0.5)$.

All in all, we gave an overview of the major dependencies of the constants in the a posteriori error estimates on the exact solution $[\lambda_0, \underline{u}_0]$ and on the material parameters. As the exact solutions are usually unknown, it is difficult to compute these constants.

4.4 Implementation

4.4.1 $\mathcal{C}o\mathcal{C}o\mathcal{S}$ – Computation of corner singularities

For the verification of the obtained results, the software package $\mathcal{C}o\mathcal{C}o\mathcal{S}$ [87] was used. $\mathcal{C}o\mathcal{C}o\mathcal{S}$ was developed by Th. Apel and U. Reichel at Chemnitz University of Technology for the computation of corner singularities for the Laplace and the linear elasticity problems. Some libraries for the further development were provided by the Sonderforschungsbereich SFB 393 of Chemnitz University of Technology [89]. In the last years, several routines were added by M. Randrianarivony, C. Pester and J. Rosam to support, for instance, the analysis of crack problems, the computation of eigenfunctions and their visualization and the calculation of stress distributions.

Furthermore, the mesh generation suggested in Section 3.2 and the computation of the residual a posteriori error estimator η_T in (3.15) were integrated for the Laplace and the linear elasticity problems. For the numerical integration, a Gauß quadrature with seven quadrature points is used for triangles, whereas the integrals over the pole elements is computed by a quadrature formula for quadrilaterals.

In the numerical tests, the eigenvalues of problem (4.26),

$$\lambda^2 m_\lambda(\underline{u}, \underline{v}) + \lambda g_\lambda(\underline{u}, \underline{v}) = k_\lambda(\underline{u}, \underline{v}),$$

were computed with the *skew-Hamiltonian implicitly restarted Arnoldi* algorithm (*shira*) [72]. The idea of the *shira* algorithm is that the discretized eigenvalue problem is linearized with a special approach and then transformed to an equivalent standard eigenvalue problem for a skew-Hamiltonian matrix. The eigenvalues and the eigenfunctions are computed separately. One method for the computation of the eigenfunctions was suggested in [45]. A simple, though more time-consuming, possibility to obtain the eigenfunctions is to perform a few steps of inverse iteration. Often, one iteration suffices to obtain a good approximation of an eigenfunction. When the discretized (matrix) eigenvalue problem reads

$$(\lambda_h^2 M_h + \lambda_h G_h - K_h)\underline{u}_h = 0$$

and λ_h is an approximation of an exact eigenvalue λ_0 of problem (4.26), then the solution $\underline{u}_h$ to the linear system of equations

$$(\lambda_h^2 M_h + \lambda_h G_h - K_h)\underline{u}_h = \underline{f}$$

with an arbitrary non-zero right-hand side $\underline{f}$ is an approximation of an eigenfunction corresponding to λ_0. The nodal basis functions in the discretization of the eigenvalue problem can be chosen so that the finite element matrices M_h, G_h and K_h are real. We used the nodal basis functions which were suggested in Remark 3.10.

Suppose that the eigenvalue λ_h and a corresponding eigenfunction $\underline{u}_h$ has been computed. To estimate the approximation error of the eigenvalues, the residuals of the adjoint problem have to be evaluated. For the Laplace problem, they do not differ from those of the primal problem. For the linear elasticity problem, it is sufficient to compute the residuals corresponding to the eigenvalue $-\bar{\lambda}_h$ because of the specific structure of the spectrum of (4.26), see Remark 4.15. The eigenfunction $\underline{u}_h^\star$ corresponding to $-\bar{\lambda}_h$ is obtained from another step of inverse iteration. The computation of the residuals is explained in detail in Section 4.4.3.

The error estimation routine is applied once to the pair $[\alpha_h, \underline{u}_h] = [\lambda_h - 1/2, \underline{u}_h]$ and, in the case of the linear elasticity problem, once to $[-1 - \bar{\alpha}_h, \underline{u}_h] = [-\bar{\lambda}_h - 1/2, \underline{u}_h]$. It returns the values η_T and $\eta_T^\star$ for all $T \in \mathcal{T}_h$ (or $\tilde{\eta}_T$ and $\tilde{\eta}_T^\star$). The mesh is refined adaptively on the basis of the error estimator for the primal problem in order to find a better approximation for the singularity exponents. A standard red-green refinement procedure [11] is used. Its realization in $\mathcal{C}o\mathcal{C}o\mathcal{S}$ is explained in [87, Section 7]. For technical reasons, only the red refinement is used for pole elements.

4.4.2 The approximate solution $\underline{u}_h$ and the condition $(\underline{u}_h, \bar{\underline{u}}_h)_H = 1$

Suppose that the approximate eigenfunction $\underline{u}_h$ has been computed. It is uniquely represented by its coefficients with respect to the nodal basis functions. Let N be the number of the nodes in $\mathcal{N}_h$ which we enumerate so that $\mathcal{N}_h = \{\mathrm{x}_i\}_{i=1}^N$. Then

$$\underline{u}_h = \sum_{i=1}^{N} \underline{u}_i \phi_{\mathrm{x}_i},$$

where $\underline{u}_i \in \mathbb{R}$ for the Laplace problem and $\underline{u}_i \in \mathbb{C}^3$ for the linear elasticity problem. When d is the number of degrees of freedom per node, then we can write $\underline{u}_i \in \mathbb{K}^d$ as $\underline{u}_i = \sum_{k=1}^d u_{ik}\mathbf{e}_k$ and

$$\underline{u}_h = \sum_{k=1}^{d} \sum_{i=1}^{N} u_{ik}\phi_{\mathrm{x}_i}\mathbf{e}_k.$$

Moreover, the total number of degrees of freedom (and the dimension of the finite element matrices) is given by $d\,N$. The finite element matrices are stored blockwise, this means, they consist of $N \times N$ blocks of the dimensions $d \times d$ each. We conclude that

$$(\underline{u}_h, \bar{\underline{u}}_h)_H = (\sum_{k=1}^{d} \sum_{i=1}^{N} u_{ik}\phi_{\mathrm{x}_i}\mathbf{e}_k, \sum_{\ell=1}^{d} \sum_{j=1}^{N} \bar{u}_{j\ell}\phi_{\mathrm{x}_i}\mathbf{e}_\ell)_H = \sum_{k,\ell=1}^{d} \sum_{i,j=1}^{N} u_{ik}u_{j\ell}(\phi_{\mathrm{x}_i}\mathbf{e}_k, \phi_{\mathrm{x}_j}\mathbf{e}_\ell)_H.$$

It is reasonable to choose an inner product for which the values $(\phi_{\mathrm{x}_i}\mathbf{e}_k, \phi_{\mathrm{x}_j}\mathbf{e}_\ell)_H$ are already known. In the generation of the finite element matrices M_h and K_h, we computed the terms $m_\lambda(\phi_{\mathrm{x}_i}\mathbf{e}_k, \phi_{\mathrm{x}_j}\mathbf{e}_\ell)$ and $k_\lambda(\phi_{\mathrm{x}_i}\mathbf{e}_k, \phi_{\mathrm{x}_j}\mathbf{e}_\ell)$ which correspond to the entries (k, ℓ) the (i, j)-th block of M_h and K_h, respectively. For this reason, we use $(\cdot,\cdot)_H = m_\lambda(\cdot,\cdot)$ in the implementation.

Suppose that $\sum_{k,\ell=1}^d \sum_{i,j=1}^N u_{ik}u_{j\ell}m_\lambda(\phi_{\mathrm{x}_i}\mathbf{e}_k, \phi_{\mathrm{x}_j}\mathbf{e}_\ell) = m_\lambda(\underline{u}_h, \bar{\underline{u}}_h) = z \neq 0$. We compute a complex root c of $c^2 = z$ and use the function $\underline{u}_h/c$ for the computation of the error estimator. The function $\tilde{\underline{u}}_h = \underline{u}_h/c$ satisfies $m_\lambda(\tilde{\underline{u}}_h, \bar{\tilde{\underline{u}}}_h) = 1/c^2 m_\lambda(\underline{u}_h, \bar{\underline{u}}_h) = 1$.

4.4.3 Computation of the residuals in spherical coordinates

Before we present the numerical results, some parts of the implementation shall be explained in detail. We motivated in Section 3.2 why spherical coordinates are used for the parametrization of the sphere. One of the reasons is that we can immediately use a triangulation routine for two-dimensional domains to split the parameter domain. When we use the mesh generation algorithm from page 42, it will never be necessary to think about a decomposition of the spherical domain $\Omega_S \subset \mathbb{R}^3$; the generated mesh will always satisfy the desired properties of the triangulation, see Section 3.2.

To clarify the meaning of the notation of Section 2.4 for the particular parametrization, recall

$$\underline{\mathbf{X}} = \underline{\mathbf{X}}(\xi_1, \xi_2, \xi_3) = \underline{\mathbf{X}}(\varphi, \theta, r) = x_1\mathbf{e}_1 + x_2\mathbf{e}_2 + x_3\mathbf{e}_3$$

for the Cartesian basis $\{\mathbf{e}_i\}_{i=1}^3$ and $x_1 = r\cos\varphi\sin\theta$, $x_2 = r\sin\varphi\sin\theta$, $x_3 = r\cos\theta$. With $\xi_1 = \varphi$, $\xi_2 = \theta$ and $\xi_3 = r$, the co- and contravariant tensor bases $\{\mathbf{g}_i\}_{i=1}^3$, $\mathbf{g}_i = \partial\underline{\mathbf{X}}/\partial\xi_i$, and $\{\mathbf{g}^j\}_{j=1}^3$, $\mathbf{g}_i \cdot \mathbf{g}^j = \delta_i^j$, can be summarized in the matrices

$$G(\varphi, \theta, r) = \left(\mathbf{g}_1\ \mathbf{g}_2\ \mathbf{g}_3 \right) = \begin{pmatrix} -r\sin\varphi\sin\theta & r\cos\varphi\cos\theta & \cos\varphi\sin\theta \\ r\cos\varphi\sin\theta & r\sin\varphi\cos\theta & \sin\varphi\sin\theta \\ 0 & -r\sin\theta & \cos\theta \end{pmatrix}$$

and

$$G^{-\mathsf{T}}(\varphi, \theta, r) = \left(\mathbf{g}^1\ \mathbf{g}^2\ \mathbf{g}^3 \right) = \frac{1}{r} \begin{pmatrix} -\sin\varphi/\sin\theta & \cos\varphi\cos\theta & r\cos\varphi\sin\theta \\ \cos\varphi/\sin\theta & \sin\varphi\cos\theta & r\sin\varphi\sin\theta \\ 0 & -\sin\theta & r\cos\theta \end{pmatrix}$$

For $r = 1$, we obtain the corresponding quantities on the unit sphere which we indicate with the index $\mathcal{S}$. In particular, let $G_{\mathcal{S}}^{ij}$ denote the entries of the matrix

$$G_{\mathcal{S}}^{-\top}(\varphi,\theta) = G^{-\top}(\varphi,\theta,1) = \left(\, \mathbf{g}_{\mathcal{S}}^1 \; \mathbf{g}_{\mathcal{S}}^2 \; \mathbf{g}_{\mathcal{S}}^3 \, \right) = \begin{pmatrix} -\sin\varphi/\sin\theta & \cos\varphi\cos\theta & \cos\varphi\sin\theta \\ \cos\varphi/\sin\theta & \sin\varphi\cos\theta & \sin\varphi\sin\theta \\ 0 & -\sin\theta & \cos\theta \end{pmatrix}.$$

We emphasize that the components of these matrices are given with respect to the Cartesian basis, for instance, $\mathbf{g}_{\mathcal{S}}^j = \sum_{i=1}^3 G_{\mathcal{S}}^{ij}\mathbf{e}_i$, $i = 1, 2, 3$. Note that $|\mathbf{g}_{\mathcal{S}}^1| = 1/\sin\theta$, $|\mathbf{g}_{\mathcal{S}}^2| = |\mathbf{g}_{\mathcal{S}}^3| = 1$ and $\mathbf{g}_{\mathcal{S}}^i \cdot \mathbf{g}_{\mathcal{S}}^j = 0$ for $i \neq j$. This is a special feature of spherical coordinates which simplifies further computations essentially.

The spherical gradient of a scalar function u reads $\nabla_{\mathcal{S}} u = \sum_{i=1}^2 \mathbf{g}_{\mathcal{S}}^i \, \partial u/\partial\xi_i$, so that

$$|\nabla_{\mathcal{S}} u|^2 = \nabla_{\mathcal{S}}\bar{u} \cdot \nabla_{\mathcal{S}} u = \left|\frac{1}{\sin\theta}\frac{\partial u}{\partial\varphi}\right|^2 + \left|\frac{\partial u}{\partial\theta}\right|^2.$$

To give a similar formula for vector functions $\underline{u}$, it is necessary to fix the basis into which $\underline{u}$ is developed. Usually the contravariant basis is used. We discussed already in Section 2.4 that it might be preferable to choose another basis at this point. For the sake of simplicity, we use the Cartesian basis $\{\mathbf{e}_1, \mathbf{e}_2, \mathbf{e}_3\}$ which is orthonormal and independent of the parameters φ, θ and r. The components u_i of $\underline{u}$ shall be defined by $\underline{u} = \sum_{i=1}^3 u_i \, \mathbf{e}_i$. Then, the second order tensor $\nabla_{\mathcal{S}}\underline{u} = \sum_{j=1}^3 \nabla_{\mathcal{S}} u_j \, \mathbf{e}_j$ satisfies

$$(\nabla_{\mathcal{S}}\underline{u})^H : (\nabla_{\mathcal{S}}\underline{u}) = \sum_{i=1}^3 \mathbf{e}_i \, \nabla_{\mathcal{S}}\bar{u}_i : \sum_{j=1}^3 \nabla_{\mathcal{S}} u_j \, \mathbf{e}_j = \sum_{i,j=1}^3 \delta_{ij} \, \nabla_{\mathcal{S}}\bar{u}_i \cdot \nabla_{\mathcal{S}} u_j = \sum_{i=1}^3 |\nabla_{\mathcal{S}} u_i|^2.$$

In the following, we will concentrate on the implementation of the elasticity problem, because the treatment of the tensors is more subtle than the theory for the Laplace problem. The material tensor A is a fourth order tensor which is usually developed four times into the Cartesian basis. Its components $a_{ijk\ell} \in \mathbb{R}$ are given implicitly by $A = \sum_{i,j,k,\ell=1}^3 a_{ijk\ell} \, \mathbf{e}_i\mathbf{e}_j\mathbf{e}_k\mathbf{e}_\ell$. Obviously, there are 81 possible coefficients to describe the material. Due to certain symmetries in the material, in particular,

$$a_{ijk\ell} = a_{jik\ell} = a_{k\ell ij}, \tag{4.35}$$

the actual number of these coefficients reduces to 21. These are the entries of the upper triangle of a symmetric positive definite matrix $\underline{A} \in \mathbb{R}^{6\times 6}$,

$$\underline{A} = \begin{pmatrix} a_{1111} & a_{1122} & a_{1133} & a_{1123} & a_{1113} & a_{1112} \\ a_{1122} & a_{2222} & a_{2233} & a_{2223} & a_{2213} & a_{2212} \\ a_{1133} & a_{2233} & a_{3333} & a_{3323} & a_{3313} & a_{3312} \\ a_{1123} & a_{2223} & a_{3323} & a_{2323} & a_{2313} & a_{2312} \\ a_{1113} & a_{2213} & a_{3313} & a_{2313} & a_{1313} & a_{1312} \\ a_{1112} & a_{2212} & a_{3312} & a_{2312} & a_{1312} & a_{1212} \end{pmatrix}.$$

A detailed discussion of the relation between the tensor operations with the fourth order tensor A and the six-dimensional matrix operations with the matrix $\underline{A}$ is given in [81, Part II, Section 15]. Usually, the number of independent material parameters can be reduced further. Isotropic material can be described by only two parameters. A material for whose description all 21 parameters are needed is copper vitriol (copper sulfate), a bluish triclinic crystal which is highly poisonous and finds application, for instance, in the therapy of thrush.

In tensor notation, relation (4.35) means that

$$\tau^\top : A = \tau : A, \qquad \tau : A : \sigma = \sigma : A : \tau \tag{4.36}$$

for all (symmetric or non-symmetric) second order tensors τ, σ. This can be easily verified by using the calculation rules from Section 2.4. For instance, let $\{\mathbf{c}_k\}_{k=1}^3$ and $\{\mathbf{d}_k\}_{k=1}^3$ be two further bases of $\mathbb{R}^3$ and suppose that $a_{jik\ell} = a_{k\ell ij}$ holds, then

$$\begin{aligned}
\tau : A : \sigma &= \sum_{p,q=1}^{3} \tau_{pq}\mathbf{c}_p\mathbf{d}_q : \sum_{i,j,k,\ell=1}^{3} a_{ijk\ell}\mathbf{e}_i\mathbf{e}_j\mathbf{e}_k\mathbf{e}_\ell : \sum_{s,t=1}^{3} \sigma_{st}\mathbf{c}_s\mathbf{d}_t \\
&= \sum_{p,q,i,j,k,h,s,t=1}^{3} \tau_{pq}a_{ijk\ell}\sigma_{st}(\mathbf{d}_q\cdot\mathbf{e}_i)(\mathbf{c}_p\cdot\mathbf{e}_j)(\mathbf{e}_\ell\cdot\mathbf{c}_s)(\mathbf{e}_k\cdot\mathbf{d}_t) \\
&= \sum_{p,q,i,j,k,h,s,t=1}^{3} \tau_{pq}a_{k\ell ij}\sigma_{st}(\mathbf{d}_q\cdot\mathbf{e}_k)(\mathbf{c}_p\cdot\mathbf{e}_\ell)(\mathbf{e}_j\cdot\mathbf{c}_s)(\mathbf{e}_i\cdot\mathbf{d}_t) \\
&= \sum_{s,t=1}^{3} \sigma_{st}\mathbf{c}_s\mathbf{d}_t : \sum_{i,j,k,\ell=1}^{3} a_{ijk\ell}\mathbf{e}_i\mathbf{e}_j\mathbf{e}_k\mathbf{e}_\ell : \sum_{p,q=1}^{3} \tau_{pq}\mathbf{c}_p\mathbf{d}_q \\
&= \sigma : A : \tau.
\end{aligned}$$

We emphasize that, from the mathematical point of view, the symmetry properties (4.35) are not needed and might be misleading. They do *not* follow from certain symmetry considerations of the stress and the strain tensors, although claimed in many books on elasticity theory including [65, 68, 81]. The symmetry assumptions in (4.35), however, are legitimate, since A is usually applied *only* to symmetric tensors, see the discussion in [75]. In this case, the relations in (4.36) coincide with $\tau : A : \sigma = \sigma : A : \tau$ for $\tau = \tau^\top$ and $\sigma = \sigma^\top$. Note that this relation is only a special case and not equivalent to (4.36) or (4.35). Using (4.36), we can write

$$A : (\tau + \tau^\top) = 2\,A : \tau \tag{4.37}$$

also for unsymmetric second order tensors τ, when we do not apply the ellipticity assumption (4.24) to the right-hand side. Otherwise, skew-symmetric tensors are extinguished by A which is a contradiction to the positive definiteness of A. Although the tensors $\varepsilon_S(\underline{u})$ and $\varepsilon_3(\underline{u})$, which were introduced in (4.17), are symmetric, one has to be careful with rash simplifications exploiting (4.37). The less cautious application of (4.37) has lead to erroneous conclusions in [7], where results similar to those in Theorem 4.11 were obtained with incorrect constants.

In (4.34), we defined the element and edge residuals $\underline{R}_T$ and $\underline{R}_E$ for the elasticity problem. Recall $\mathbf{e}_i \cdot \mathbf{g}_S^j = G_S^{ij}$. Due to (4.17), (4.18) and (4.37), the element residual $\underline{R}_T$ can be written as

$$\begin{aligned}
\underline{R}_T &= \{-\nabla_S\cdot\sigma_S(\alpha_h, \underline{u}_h) - (\alpha_h - 1)\,\underline{\mathbf{x}}\cdot\sigma_S(\alpha_h,\underline{u}_h)\}|_T \\
&= \Big(-\nabla_S - (\alpha_h - 1)\,\underline{\mathbf{x}}\Big)\cdot\Big(A : \Big[\varepsilon_S(\underline{u}_h) + \alpha_h\varepsilon_3(\underline{u}_h)\Big]\Big) \\
&= \Big(-\nabla_S - (\alpha_h - 1)\,\underline{\mathbf{x}}\Big)\cdot\Big(A : \Big[\nabla_S\underline{u}_h + \alpha_h\,\underline{\mathbf{x}}\underline{u}_h\Big]\Big)
\end{aligned}$$

$$
\begin{aligned}
&= \Big(-\sum_{q=1}^{2} \mathbf{g}_{\mathcal{S}}^{q} \frac{\partial}{\partial \xi_q} - (\alpha_h - 1)\,\mathbf{g}^3\Big) \cdot \\
&\qquad \cdot \Big(\Big[\sum_{i,j,k,\ell=1}^{3} a_{ijk\ell}\, \mathbf{e}_i \mathbf{e}_j \mathbf{e}_k \mathbf{e}_\ell\Big] : \Big[\sum_{p=1}^{2}\sum_{n=1}^{3} \frac{\partial u_n}{\partial \xi_p} \mathbf{g}_{\mathcal{S}}^{p} \mathbf{e}_n + \alpha_h \sum_{n=1}^{3} u_n\, \mathbf{g}^3 \mathbf{e}_n\Big]\Big) \\
&= \Big(-\sum_{q=1}^{2} \mathbf{g}_{\mathcal{S}}^{q} \frac{\partial}{\partial \xi_q} - (\alpha_h - 1)\,\mathbf{g}^3\Big) \cdot \\
&\qquad \cdot \Big(\sum_{i,j,k,\ell=1}^{3} \sum_{n=1}^{3} a_{ijk\ell} \Big\{\sum_{p=1}^{2} \frac{\partial u_n}{\partial \xi_p} (\mathbf{e}_\ell \cdot \mathbf{g}_{\mathcal{S}}^{p})\,(\mathbf{e}_k \cdot \mathbf{e}_n) + \alpha_h\, u_n\, (\mathbf{e}_\ell \cdot \mathbf{g}^3)\,(\mathbf{e}_k \cdot \mathbf{e}_n)\Big\} \mathbf{e}_i \mathbf{e}_j \\
&= \Big(-\sum_{q=1}^{2} \mathbf{g}_{\mathcal{S}}^{q} \frac{\partial}{\partial \xi_p} - (\alpha_h - 1)\,\mathbf{g}^3\Big) \cdot \Big(\sum_{i,j,k,\ell=1}^{3} a_{ijk\ell} \Big\{\sum_{p=1}^{2} \frac{\partial u_k}{\partial \xi_p} G_{\mathcal{S}}^{\ell p} + \alpha_h\, u_k\, G_{\mathcal{S}}^{\ell 3}\Big\} \mathbf{e}_i \mathbf{e}_j\Big) \\
&= -\sum_{i,j,k,\ell=1}^{3} a_{ijk\ell} \sum_{q=1}^{2} \frac{\partial}{\partial \xi_q} \Big\{\sum_{p=1}^{2} \frac{\partial u_k}{\partial \xi_p} G_{\mathcal{S}}^{\ell p} + \alpha_h\, u_k\, G_{\mathcal{S}}^{\ell 3}\Big\} (\mathbf{g}_{\mathcal{S}}^{q} \cdot \mathbf{e}_i)\, \mathbf{e}_j \\
&\qquad - (\alpha_h - 1) \sum_{i,j,k,\ell=1}^{3} a_{ijk\ell} \Big\{\sum_{p=1}^{2} \frac{\partial u_k}{\partial \xi_p} G_{\mathcal{S}}^{\ell p} + \alpha_h\, u_k\, G_{\mathcal{S}}^{\ell 3}\Big\} (\mathbf{g}^3 \cdot \mathbf{e}_i)\, \mathbf{e}_j \\
&= -\sum_{i,j,k,\ell=1}^{3} a_{ijk\ell} \sum_{q=1}^{2} \frac{\partial}{\partial \xi_q} \Big\{\sum_{p=1}^{2} \frac{\partial u_k}{\partial \xi_p} G_{\mathcal{S}}^{\ell p} + \alpha_h\, u_k\, G_{\mathcal{S}}^{\ell 3}\Big\} G_{\mathcal{S}}^{iq}\, \mathbf{e}_j \\
&\qquad - (\alpha_h - 1) \sum_{i,j,k,\ell=1}^{3} a_{ijk\ell} \Big\{\sum_{p=1}^{2} \frac{\partial u_k}{\partial \xi_p} G_{\mathcal{S}}^{\ell p} + \alpha_h\, u_k\, G_{\mathcal{S}}^{\ell 3}\Big\} G_{\mathcal{S}}^{i3}\, \mathbf{e}_j.
\end{aligned}
$$

According to (4.35), the indices i and j can be exchanged, so that the components of $\underline{R}_T = \sum_{i=1}^{3} R_{T,i} \mathbf{e}_i$ are given by

$$
\begin{aligned}
R_{T,i} &= -\sum_{j,k,\ell=1}^{3} a_{ijk\ell} \sum_{q=1}^{2} \Big\{\sum_{p=1}^{2} \frac{\partial^2 u_k}{\partial \xi_p \partial \xi_q} G_{\mathcal{S}}^{\ell p} + \frac{\partial u_k}{\partial \xi_p} \frac{\partial G_{\mathcal{S}}^{\ell p}}{\partial \xi_q} + \alpha_h \frac{\partial u_k}{\partial \xi_q} G_{\mathcal{S}}^{\ell 3} + \alpha_h\, u_k \frac{\partial G_{\mathcal{S}}^{\ell 3}}{\partial \xi_q}\Big\} G_{\mathcal{S}}^{jq} \\
&\qquad - (\alpha_h - 1) \sum_{j,k,\ell=1}^{3} a_{ijk\ell} \Big\{\sum_{p=1}^{2} \frac{\partial u_k}{\partial \xi_p} G_{\mathcal{S}}^{\ell p} + \alpha_h\, u_k\, G_{\mathcal{S}}^{\ell 3}\Big\} G_{\mathcal{S}}^{j3}.
\end{aligned}
$$

Furthermore, an edge E is parametrized in the form $\{(\underline{\mathbf{x}}(\xi_1(t), \xi_2(t)) \mid 0 < t < 1\}$. The normal vector $\mathbf{n}_E$ is given by $\mathbf{n}_E = \tilde{\mathbf{n}}_E / |\tilde{\mathbf{n}}_E|$, where $\tilde{\mathbf{n}}_E = \dot{\xi}_1 \mathbf{g}_{\mathcal{S}}^{2} - \dot{\xi}_2 \mathbf{g}_{\mathcal{S}}^{1}$ and

$$
|\tilde{\mathbf{n}}_E| = |\dot{\xi}_1 \mathbf{g}_{\mathcal{S}}^{2} - \dot{\xi}_2 \mathbf{g}_{\mathcal{S}}^{1}|^2 = (\dot{\xi}_1 \mathbf{g}_{\mathcal{S}}^{2} - \dot{\xi}_2 \mathbf{g}_{\mathcal{S}}^{1}) \cdot (\dot{\xi}_1 \mathbf{g}_{\mathcal{S}}^{2} - \dot{\xi}_2 \mathbf{g}_{\mathcal{S}}^{1}) = \dot{\varphi}^2 + \frac{1}{\sin^2 \theta} \dot{\theta}^2.
$$

Hence, $1/|\tilde{\mathbf{n}}_E| = \sin\theta / \sqrt{\sin^2 \theta \dot{\varphi}^2 + \dot{\theta}^2}$. With the same arguments as for $\underline{R}_T$, we obtain that

$$
\begin{aligned}
\underline{R}_E &= \lfloor \sigma_{\mathcal{S}}(\alpha_h, \underline{u}_h) \cdot \mathbf{n}_E \rfloor_E = \Big\lfloor \sigma_{\mathcal{S}}(\alpha_h, \underline{u}_h) \cdot \frac{\dot{\xi}_1 \mathbf{g}_{\mathcal{S}}^{2} - \dot{\xi}_2 \mathbf{g}_{\mathcal{S}}^{1}}{|\dot{\xi}_1 \mathbf{g}_{\mathcal{S}}^{2} - \dot{\xi}_2 \mathbf{g}_{\mathcal{S}}^{1}|} \Big\rfloor_E \\
&= \Big\lfloor \sum_{i,j,k,\ell=1}^{3} a_{ijk\ell} \Big\{\sum_{p=1}^{2} \frac{\partial u_k}{\partial \xi_p} G_{\mathcal{S}}^{\ell p} + \alpha_h u_k G^{\ell 3}\Big\} \mathbf{e}_i \mathbf{e}_j \cdot \frac{\dot{\xi}_1 \mathbf{g}_{\mathcal{S}}^{2} - \dot{\xi}_2 \mathbf{g}_{\mathcal{S}}^{1}}{|\dot{\xi}_1 \mathbf{g}_{\mathcal{S}}^{2} - \dot{\xi}_2 \mathbf{g}_{\mathcal{S}}^{1}|} \Big\rfloor_E
\end{aligned}
$$

$$= \frac{1}{|\dot{\xi}_1 \mathbf{g}_{\mathcal{S}}^2 - \dot{\xi}_2 \mathbf{g}_{\mathcal{S}}^1|} \sum_{i,j,k,\ell=1}^{3} a_{ijk\ell} \left[\sum_{p=1}^{2} \frac{\partial u_k}{\partial \xi_p} G_{\mathcal{S}}^{\ell p} + \alpha_h u_k G^{\ell 3} \right]_E (\dot{\xi}_1 G_{\mathcal{S}}^{j2} - \dot{\xi}_2 G_{\mathcal{S}}^{j1})\, \mathbf{e}_i$$

The components of $\underline{R}_E = \sum_{i=1}^{3} R_{E,i}\mathbf{e}_i$ are given by

$$R_{E,i} = \frac{\sin\theta}{\sqrt{\sin^2\theta\dot{\varphi}^2 + \dot{\theta}^2}} \sum_{j,k,\ell=1}^{3} a_{ijk\ell} \left[\sum_{p=1}^{2} \frac{\partial u_k}{\partial \xi_p} G_{\mathcal{S}}^{\ell p} + \alpha_h u_k G^{\ell 3} \right]_E (\dot{\varphi} G_{\mathcal{S}}^{j2} - \dot{\theta} G_{\mathcal{S}}^{j1})$$

The components u_k of the function $\underline{u}_h = \sum_{k=1}^{3} u_k \mathbf{e}_k$ are linear combinations of piecewise linear or bilinear nodal basis functions, see Remark 3.10. That is, $u_k|_T = \sum_{n=1}^{3} u_{kn}\phi_{\mathrm{x}_n,T}$ with certain constants u_{kn} which are independent of φ and θ. This means that we actually compute the partial derivatives of the nodal basis functions. In particular, for non-pole elements, all second derivatives of the nodal basis functions vanish; for pole elements, the mixed derivatives remain, but are constant. Furthermore, the terms $\partial G_{\mathcal{S}}^{\ell p}/\partial \xi_q$, $\ell, p = 1,2,3$, $q = 1,2$, are needed for the implementation of the residuals. They are summarized in the matrices

$$\frac{\partial}{\partial\varphi} G_{\mathcal{S}}^{-\top} = \frac{\partial}{\partial\varphi} \begin{pmatrix} G_{\mathcal{S}}^{11} & G_{\mathcal{S}}^{12} & G_{\mathcal{S}}^{13} \\ G_{\mathcal{S}}^{21} & G_{\mathcal{S}}^{22} & G_{\mathcal{S}}^{23} \\ G_{\mathcal{S}}^{31} & G_{\mathcal{S}}^{31} & G_{\mathcal{S}}^{33} \end{pmatrix} = \begin{pmatrix} -\cos\varphi/\sin\theta & -\sin\varphi\cos\theta & -\sin\varphi\sin\theta \\ -\sin\varphi/\sin\theta & \cos\varphi\cos\theta & \cos\varphi\sin\theta \\ 0 & 0 & 0 \end{pmatrix}$$

and

$$\frac{\partial}{\partial\theta} G_{\mathcal{S}}^{-\top} = \frac{\partial}{\partial\theta} \begin{pmatrix} G_{\mathcal{S}}^{11} & G_{\mathcal{S}}^{12} & G_{\mathcal{S}}^{13} \\ G_{\mathcal{S}}^{21} & G_{\mathcal{S}}^{22} & G_{\mathcal{S}}^{23} \\ G_{\mathcal{S}}^{31} & G_{\mathcal{S}}^{31} & G_{\mathcal{S}}^{33} \end{pmatrix} = \begin{pmatrix} \sin\varphi\cos\theta/\sin^2\theta & -\cos\varphi\sin\theta & \cos\varphi\cos\theta \\ -\cos\varphi\cos\theta/\sin^2\theta & -\sin\varphi\sin\theta & \sin\varphi\cos\theta \\ 0 & -\cos\theta & -\sin\theta \end{pmatrix}.$$

Provided that the material parameters $a_{ijk\ell}$ are known, all information for the implementation of the residual error estimator is given in the above formulae.

4.5 Numerical results

In general, the exact solutions to the eigenvalue problems (4.8) and (4.22) are not known. In special cases, it is possible to derive transcendental equations [97, 84] from which the singularity exponents can computed, for example, with a Newton method. Usually, however, the eigenvalue problems have to be discretized and to be solved, for example, with an Arnoldi or Lanczos algorithm. The algorithms used in the program $\mathcal{C}o\mathcal{C}o\mathcal{S}$ [87] for the computation of the corner singularities were shortly described and compared with respect to their efficiency in [5], see also [94] and references therein.

In the following subsections, we present the results of some numerical experiments, where we computed the error estimators $\eta = \{\sum_{T\in\mathcal{T}_h} \eta_T^2\}^{1/2}$ for the primal eigenvalue problem and $\eta^\star = \{\sum_{T\in\mathcal{T}_h} \eta_T^{\star 2}\}^{1/2}$ for the adjoint eigenvalue problem, compare the definitions (3.15) and (3.32) on the pages 49 and 57. According to the results known from a posteriori error estimates, see, for instance, [7], it makes sense to scale the error estimators by the factor $1/|\alpha_0|^2$, when a good approximation α_0 to the exact eigenvalue is known. For each of the example domains, we display the development of these (scaled) errors once with respect to the eigenvalue problem (4.8) for the Laplace-Beltrami operator and once with respect to quadratic eigenvalue problem (4.22) which is associated with the linear elasticity problem in a domain with corners.

In each case, we start with a rather coarse initial mesh and refine it adaptively based on the residual errors η_T of the primal problem, see the introduction of Section 4.4. The values of $\eta_T^\star$ do not influence the refinement process. It can therefore be expected that, concerning the linear elasticity problem, the error estimator $\eta\,\eta^\star$ for the eigenvalues overestimates the error $|\alpha_0 - \alpha_h|$ by a larger constant than concerning the Laplace-Beltrami operator, where the primal and adjoint problems coincide so that $\eta_T = \eta_T^\star$ for all elements T of the triangulation $\mathcal{T}_h$. Another reason for possibly larger constants in the estimates concerning the linear elasticity problem is that they depend on the material parameters and that the associated eigenvalue problem is quadratic in the eigenvalue, which requires a more precise approximation of the exact eigenvalue according to Remark 2.11. We used isotropic material with the Poisson ratio $\nu = 0.2$ during all tests concerning the linear elasticity problem.

For the derivation of a lower error bound, we used element and edge residuals which are projected into the space of constant functions. According to Remark 3.26 and Corollary 3.16, the unprojected residuals define a reliable and efficient error estimator $\tilde{\eta} = \{\tilde{\eta}_T^2\}^{1/2}$ as well. The numerical experiments were performed for both the projected and the unprojected residuals. As no significant differences concerning the development of the error or the constants could be observed during the tests, we will present in the following only the results with respect to the projected residuals (the approximation errors ε_T are neglected).

4.5.1 Notch with variable angle ξ

Let Ω be a notch with the angle $\xi > \pi$ as illustrated in Figure 4.1. We choose a point on the edge, in whose neighborhood the Laplace problem or the linear elasticity problem is to be solved, and intersect the notch with a ball of radius 1 which is centered at the marked point. The resulting intersection domain Ω_S is a zone on the unit sphere which is bounded by two geodesic lines. This spherical domain is parametrized by the spherical angles $\varphi \in (0, \xi)$ and $\theta \in [0, \pi]$. The corresponding parameter domain is shown in Figure 4.1, where the bold lines represent the boundary of Ω_S. In the numerical tests, we chose a 300°-notch, that is $\xi = \frac{5}{3}\pi$.

Laplace problem

We consider the eigenvalue problem (4.8),

$$-\Delta_S u = \alpha(\alpha + 1)u,$$

for the notch once with pure Dirichlet boundary conditions ($\Gamma_D = \partial\Omega_S$) and once with mixed boundary conditions ($\Gamma_D = \{\underline{\mathbf{x}}(\varphi, \theta) \in \partial\Omega_S \mid \varphi = 0\}$, $\Gamma_N = \{\underline{\mathbf{x}}(\varphi, \theta) \in \partial\Omega_S \mid \varphi = \xi\}$).

The exact eigenvalues are known for the cross section of the notch which is a two-dimensional domain with a concave corner of the angle ξ. In the case of pure Dirichlet or Neumann boundary conditions, the singularity exponents α can be expressed by $k\frac{\pi}{\xi}$, $k \in \mathbb{N}$ ($k > 0$ for the Dirichlet problem), and in the case of mixed boundary conditions by $(k + \frac{1}{2})\frac{\pi}{\xi}$, $k \geq 0$. For the three-dimensional notch, additional singularity exponents occur: each value $\alpha = k\frac{\pi}{\xi} + n$ (or $(k + \frac{1}{2})\frac{\pi}{\xi} + n$) with $n \in \mathbb{N}$ is an eigenvalue of problem (4.8) as well. The eigenpairs in $3D$ which have no counterpart in $2D$ are called *shadows* or *shadow terms* [83, 25].

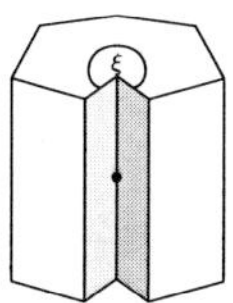

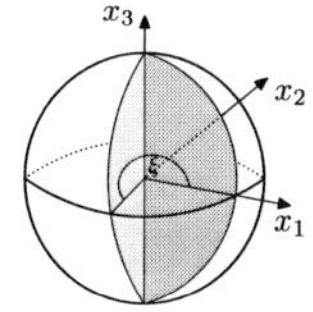

Figure 4.1: Notch with variable angle ξ, its intersection with the unit sphere and the corresponding parameter domain (in the numerical tests: $\xi = \frac{5}{3}\pi$)

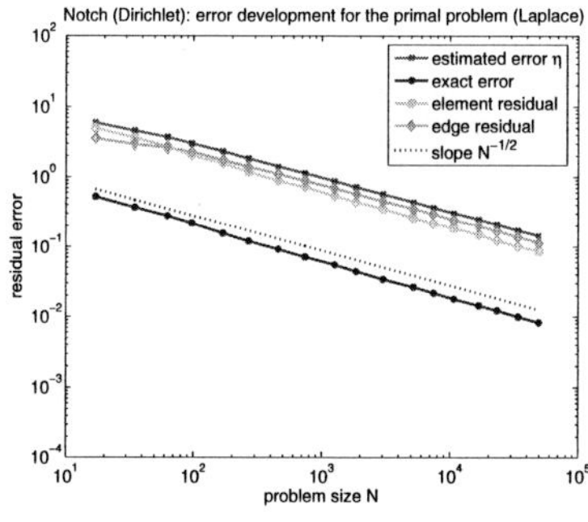

Figure 4.2: Notch – development of the projected residuals and the error estimator η during an adaptive refinement for problem (4.8) for $\alpha \approx 0.6$

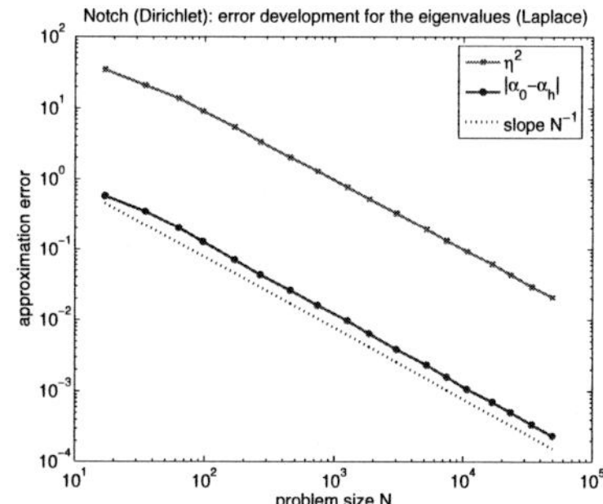

Figure 4.3: Notch – development of $|\alpha_0 - \alpha_h|$ and the error estimator η^2 during an adaptive refinement for problem (4.8) for $\alpha \approx 0.6$

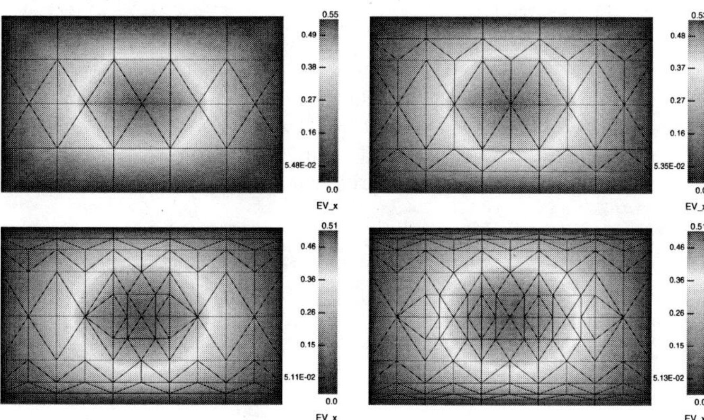

Figure 4.4: Notch – initial mesh and adaptively refined meshes with appropriate approximate solutions to problem (4.8) for $\alpha \approx 0.6$

Let us first consider the pure Dirichlet problem. Given an exponent α, there is a corresponding eigenfunction of problem (4.8) of the form $u = (\sin\theta)^\alpha \sin(\alpha\varphi)$. This is easily seen from

$$\begin{aligned}
\Delta_S u &= \frac{1}{\sin^2\theta}\frac{\partial^2 u}{\partial\varphi^2} + \frac{1}{\sin\theta}\frac{\partial}{\partial\theta}\Big(\sin\theta\frac{\partial u}{\partial\theta}\Big)\\
&= -\frac{1}{\sin^2\theta}\alpha^2(\sin\theta)^\alpha\sin(\alpha\varphi) + \frac{1}{\sin\theta}\frac{\partial}{\partial\theta}\Big(\alpha(\sin\theta)^\alpha\cos\theta\sin(\alpha\varphi)\Big)\\
&= -\frac{1}{\sin^2\theta}\alpha^2 u + \frac{1}{\sin\theta}\alpha\Big(\alpha(\sin\theta)^{\alpha-1}\cos^2\theta - (\sin\theta)^\alpha\sin\theta\Big)\sin(\alpha\varphi)\\
&= -\frac{1}{\sin^2\theta}\alpha^2 u + \frac{1}{\sin^2\theta}\alpha^2 u\cos^2\theta - \alpha u\\
&= -\alpha(\alpha+1)u
\end{aligned}$$

and $u = 0$ on $\Gamma_D = \partial\Omega_S$.

We computed an error estimator for the eigenpairs of problem (4.8). According to Section 4.2, the appropriate element and edge residuals for the discretized eigenvalue problem are given by

$$R_T = -\Delta_S u_h - \alpha(\alpha+1)_h u_h, \qquad R_E = \lfloor \nabla_S u_h \cdot \mathbf{n}_E \rfloor_E, \qquad R_N = \nabla_S u_h \cdot \mathbf{n}_E.$$

We restrict the numerical experiments to $k = 1$ and $n = 0$, that is $\alpha_0 = \frac{5}{3} = 0.6$. A corresponding eigenfunction is given by $u = (\sin\theta)^{0.6}\sin(0.6\varphi)$. The development of the residual terms

$$\Big\{\sum_{T\in\mathcal{T}_h} h_{+,T}^2 \|\pi_{0,T}R_T\|_{0,T}^2\Big\}^{1/2}, \quad \Big\{\sum_{T\in\mathcal{T}_h}\sum_{E\in\mathcal{E}(T)\cap\mathcal{E}_{h,\Omega_S}} \lceil E\rceil/\lceil T\rceil h_{+,T}^2\|\pi_{0,E}R_E\|_{0,E}^2\Big\}^{1/2}$$

and the corresponding error estimator $\eta = \{\sum_T \eta_T^2\}^{1/2}$ as well as the exact error $|\alpha_0 - \alpha_h| + \lceil u_0 - u_h\rceil_{1,\Omega_S}$ in the course of a standard red-green adaptive refinement are displayed in Figure 4.2. The slope of the dotted line corresponds to $N^{-1/2}$ which is the expected convergence order (N is the number of degrees of freedom an behaves like h^{-2}, where h is the global mesh size). Figure 4.3 shows the appropriate development of η^2 in comparison with the slope of the exact error $|\alpha_0 - \alpha_h|$ for $\alpha_0 = 0.6$. The dotted line has the slope $N^{-1} \sim h^2$.

In addition, Figure 4.4 shows a sequence of meshes which were created during the adaptive refinement process. In the background of each picture, the computed approximate solution u_h corresponding to the eigenvalue $\alpha_h \approx 0.6$ is printed. As expected, the mesh is particularly refined towards the poles, where the singularity is strongest.

Furthermore, we considered the eigenvalue problem (4.8) for the notch with mixed boundary conditions: homogeneous Dirichlet boundary condition at the line $\varphi \equiv 0$ and homogeneous Neumann boundary conditions at the line $\varphi \equiv \xi = \frac{5}{6}\pi$. One exact eigenvalue is known to be $\alpha = 0.3$. The appropriate results are displayed in Figures 4.5–4.7.

Elasticity problem

We consider the eigenvalue problem (4.22),

$$\int_{\Omega_S} \sigma_S(\alpha,\underline{u}) : \Big[(\alpha+1)\,\varepsilon_3(\underline{v}) - \varepsilon_S(\underline{v})\Big]\,dS \;=\; 0 \quad \forall v \in V$$

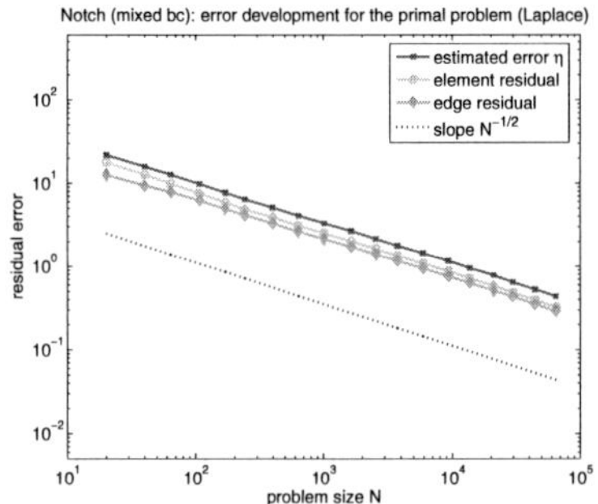

Figure 4.5: Notch – development of the projected residuals and the error estimator η during an adaptive refinement for problem (4.8) for $\alpha \approx 0.3$

Figure 4.6: Notch – development of $|\alpha_0 - \alpha_h|$ and the error estimator η^2 during an adaptive refinement for problem (4.8) for $\alpha \approx 0.3$

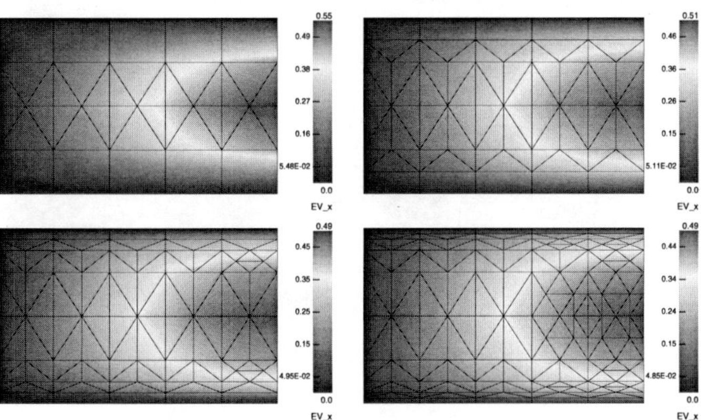

Figure 4.7: Notch – initial mesh and adaptively refined meshes with appropriate approximate solutions to problem (4.8) for $\alpha \approx 0.3$

for the notch. The exact solutions to problem (4.22) are usually unknown. A few solutions, however, can be constructed with the knowledge of the exact solutions to problem (4.8); for example, the vector function $\underline{u} = \sum_{i=1}^{3} u_i \mathbf{e}_i$ with $u_1 = u_2 = 0$ and $u_3 = (\sin\theta)^\alpha \sin(\alpha\varphi)$, is an eigenfunction of (4.22) corresponding to the singularity exponent $\alpha = \pi/\xi$ which satisfies homogeneous Dirichlet boundary conditions on $\partial\Omega_S$. We remark that the function $u_3 = (\sin\theta)^\alpha \sin(\alpha\varphi)$ corresponds to an edge singularity.

Recall from Section 4.3, equation (4.34), that the element and edge residuals corresponding to the discretized eigenvalue problem are given by

$$\begin{aligned}\underline{R}_T &= \{-\nabla_S \cdot \sigma_S(\alpha_h, \underline{u}_h) + (\alpha_h - 1)\, \underline{\mathbf{x}} \cdot \sigma_S(\alpha_h, \underline{u}_h)\}|_T,\\ \underline{R}_E &:= \lfloor \sigma_S(\alpha_h, \underline{u}_h) \cdot \mathbf{n}_E \rfloor_E, \quad \underline{R}_N := \sigma_S(\alpha_h, \underline{u}_h) \cdot \mathbf{n}_E.\end{aligned}$$

The numerical tests were performed for the approximation of the eigenvalue $\alpha_0 = 0.6$. Figure 4.8 shows the development of the residual terms

$$\Big\{\sum_{T\in\mathcal{T}_h} h_{+,T}^2 \|\pi_{0,T}\underline{R}_T\|_{0,T}^2\Big\}^{1/2}, \qquad \Big\{\sum_{T\in\mathcal{T}_h} \sum_{E\in\mathcal{E}(T)\cap\mathcal{E}_{h,\Omega_S}} \frac{\lceil E\rceil}{\lceil T\rceil} h_{+,T}^2 \|\pi_{0,E}\underline{R}_E\|_{0,E}^2\Big\}^{1/2},$$

the corresponding error estimator $\eta = \{\sum_{T\in\mathcal{T}_h} \eta_T^2\}^{1/2}$ and the exact error $|\alpha_0 - \alpha_h| + \lceil u_0 - u_h \rceil_{1,\Omega_S}$ during the adaptive refinement of the mesh. The dotted line has the slope of the function $N^{-1/2}$, where N is the dimension of the discretized problem (the number of degrees of freedom), that is $N^{-1/2} \sim h$.

For a separate estimate for the eigenvalues of problem (4.22), the term $\eta^\star = \{\sum_{T\in\mathcal{T}_h} \eta_T^{\star 2}\}^{1/2}$ has to be evaluated. The residuals for the dual problem are obtained by computing the eigenfunction $\underline{u}_h^\star$ corresponding to the eigenvalue $-1 - \bar{\alpha}_h \approx -1.6$. The results are displayed in Figure 4.9, where the dotted line has the slope of the function $N^{-1} \sim h^2$.

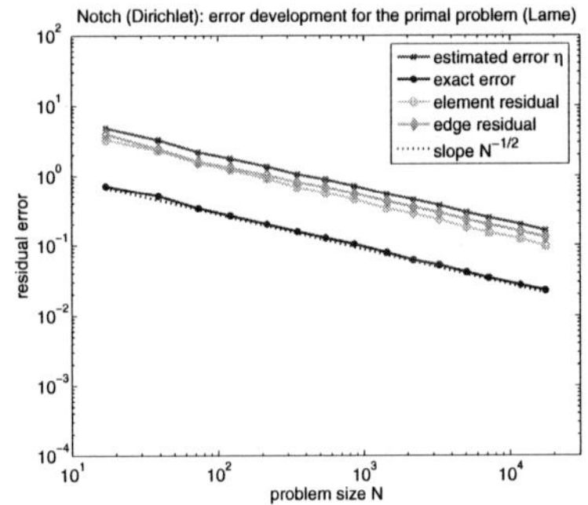

Figure 4.8: Notch – development of the projected residuals and the error estimator η during an adaptive refinement for problem (4.22) for $\alpha \approx 0.6$

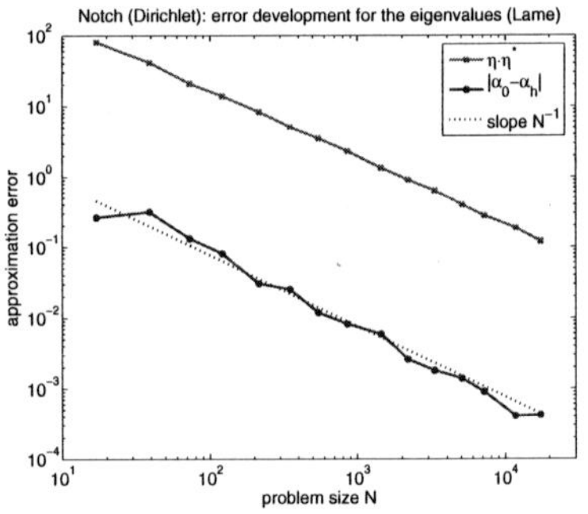

Figure 4.9: Notch – development of $|\alpha_0 - \alpha_h|$ and the error estimator $\eta\,\eta^\star$ during an adaptive refinement for problem (4.22) for $\alpha \approx 0.6$

Figure 4.10 shows the representation of the components of the solution $\underline{u}_h$ and $\underline{u}_h^\star$ to the primal and adjoint eigenvalue problems in the parameter domain corresponding to the

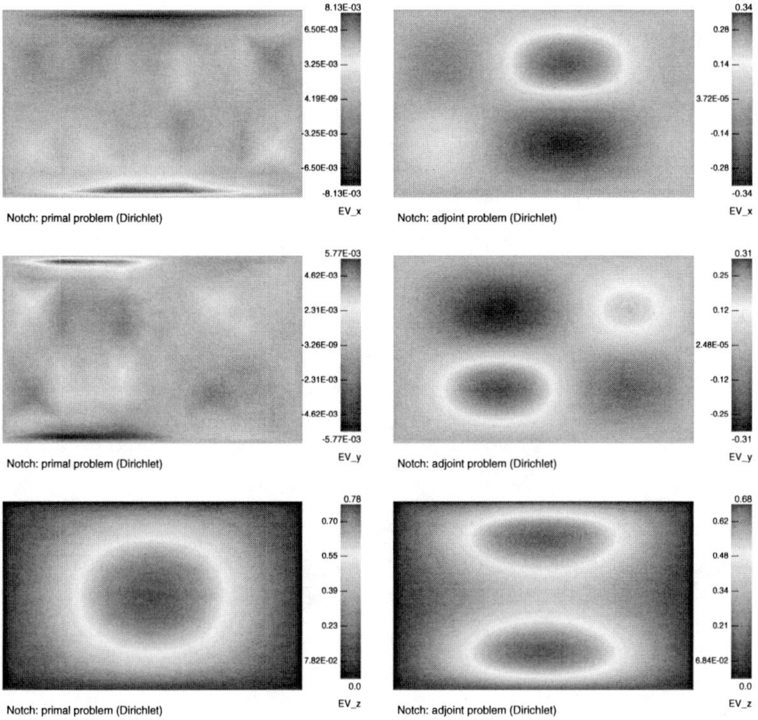

Figure 4.10: Notch – the components of the solution solutions $\underline{u}_h$ and $\underline{u}_h^\star$ to (4.22) and the corresponding adjoint eigenvalue problem for $\alpha \approx 0.6$

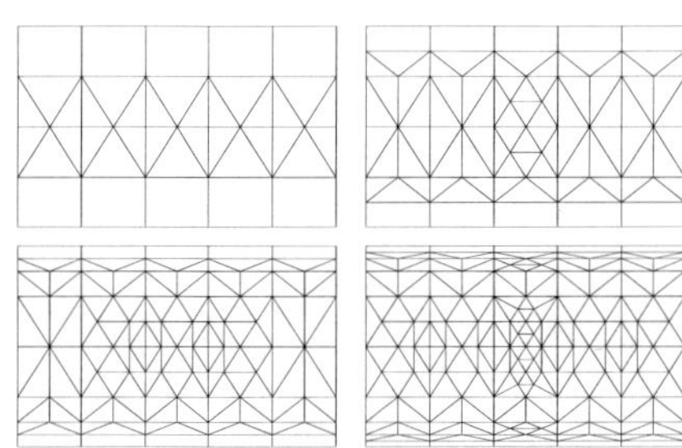

Figure 4.11: Notch – initial mesh and adaptively refined meshes for problem (4.22) for $\alpha \approx 0.6$

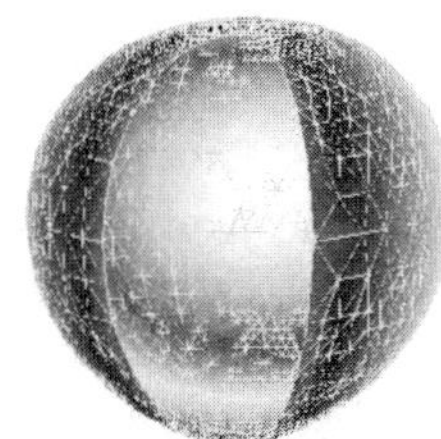

Figure 4.12: Notch – the distorted spherical domain $\underline{\mathbf{x}} + \underline{u}(\underline{\mathbf{x}})$ for problem (4.22) with $\alpha \approx 0.6$

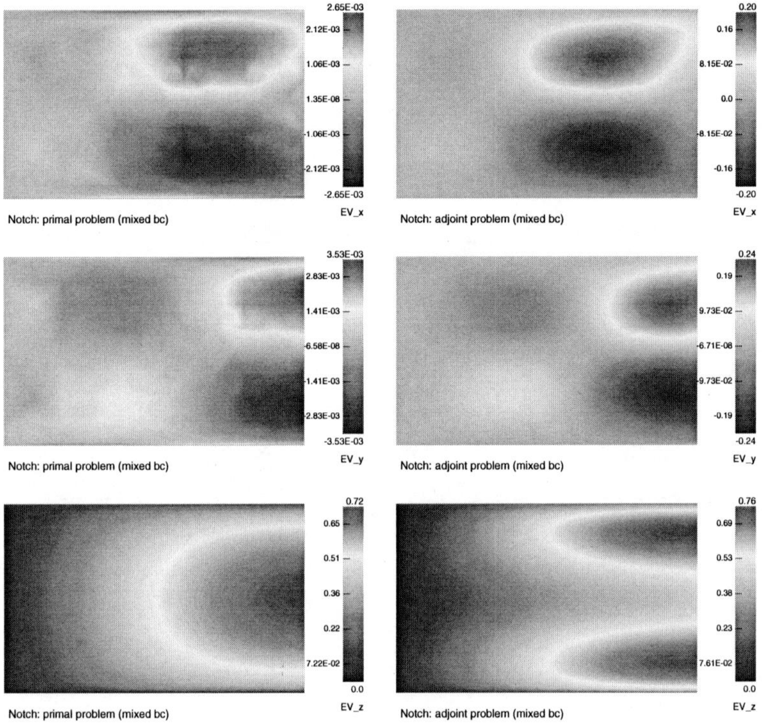

Figure 4.13: Notch – the components of the solution solutions $\underline{u}_h$ and $\underline{u}_h^\star$ to problem (4.22) and the corresponding adjoint eigenvalue problem for $\alpha \approx 0.3$

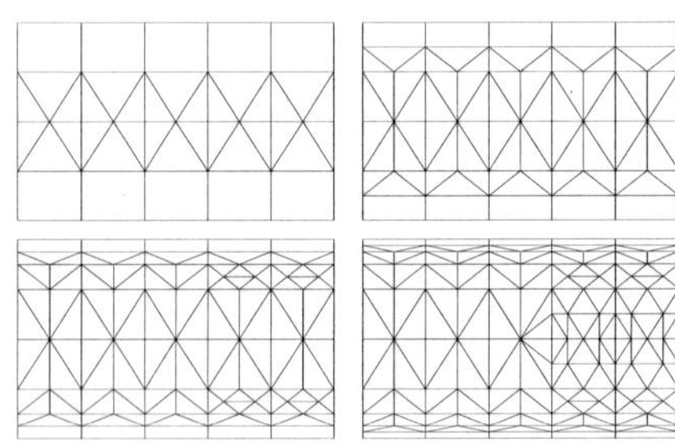

Figure 4.14: Notch – initial mesh and adaptively refined meshes for problem (4.22) for $\alpha \approx 0.3$

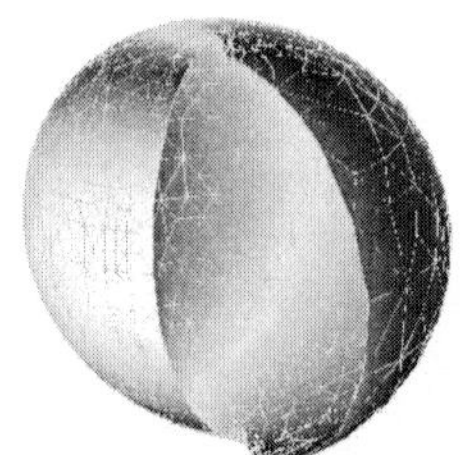

Figure 4.15: Notch – the distorted spherical domain $\underline{\mathbf{x}} + \underline{u}(\underline{\mathbf{x}})$ for problem (4.22) with $\alpha \approx 0.3$

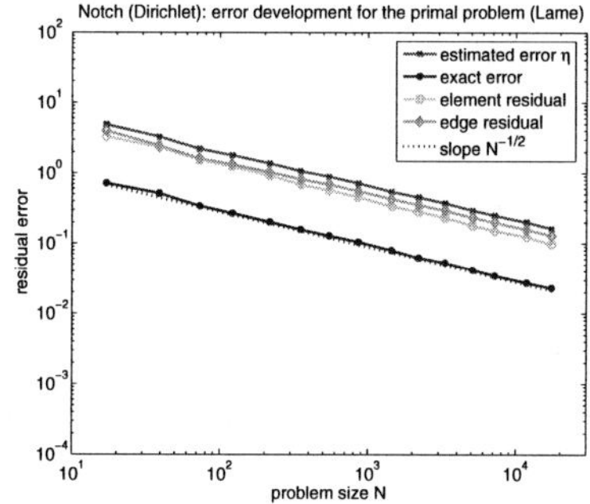

Figure 4.16: Notch – development of the projected residuals and the error estimator η during an adaptive refinement for problem (4.22) for $\alpha \approx 0.3$

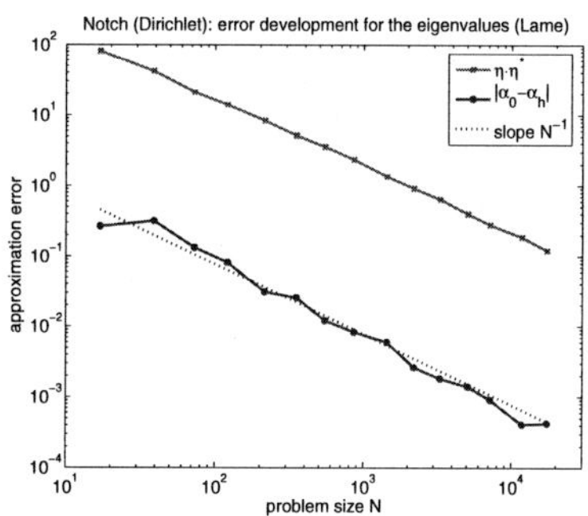

Figure 4.17: Notch – development of $|\alpha_0 - \alpha_h|$ and the error estimator $\eta\,\eta^\star$ during an adaptive refinement for problem (4.22) for $\alpha \approx 0.3$

eigenvalue $\alpha_h \approx 0.6$. (The first two components of $\underline{u}_h$ are almost zero.) The initial mesh that was used for these tests is displayed in Figure 4.11 accompanied by the meshes of the next three adaptive refinement steps. We started with a fixed number of 5 divisions into the φ-direction and 4 divisions into the θ-direction. As expected, the error is largest near the poles, which is why the pole elements are refined in each step. In addition, the distorted spherical domain, where the computed displacements $u(\underline{\mathbf{x}})$ (of the primal problem) are added to the points $\underline{\mathbf{x}} \in \Omega_S$, is displayed in Figure 4.12.

Furthermore, we considered the eigenvalue problem (4.22) for the notch with mixed boundary conditions: homogeneous Dirichlet boundary condition at the line $\varphi \equiv 0$ and homogeneous Neumann boundary conditions at the line $\varphi \equiv \xi = \frac{5}{6}\pi$. The eigenvalues α with $\mathrm{Re}(\alpha) < 1$ are $\alpha = 0.3$ and $\alpha = 0.9$. Each of them is a triple eigenvalue. Although the theory was proven only for simple eigenvalues, this example shows that it is not a necessary condition that α_0 is a simple eigenvalue The appropriate results for $\alpha_0 = 0.3$ are displayed in Figures 4.13–4.17.

4.5.2 Fichera corner

The Fichera corner, named after G. Fichera (1922–1996), is a 7/8-th cube and the prototype of a domain, where corner and edge singularities interact. Let Ω be the infinite conical domain produced by the 7/8-th cube $(-a,a)^3 \setminus [0,a]^3$ for $a \to \infty$. The intersection with the unit ball centered at the corner $(0,0,0)$ results in a 7/8-th ball as demonstrated in Figure 4.18. The surface Ω_S is parametrized by the spherical angles φ and θ. For technical reasons (symmetry considerations), we rotate the coordinate system so that the parameter domain spanned by φ and θ is the polygon

$$\tilde{\Omega}_S = [0,2\pi) \times [0,\pi] \setminus \{(\varphi,\theta) \in \mathbb{R}^2 \mid \varphi \in [\frac{3}{4}\pi, \frac{5}{4}\pi], \theta \in [0, \frac{1}{2}\pi]\},$$

see Figure 4.18. The bold lines in the picture correspond the boundary of Ω_S. The dashed lines represent periodic boundary conditions, cf. Section 3.2.

We remark that the angle ξ at the corner can also be varied, see [87] for further results. Here, we restrict the numerical tests to $\xi = \pi/2$ and Dirichlet boundary conditions ($\Gamma_D = \{\underline{\mathbf{x}}(\varphi,\theta) \in \partial\Omega_S \mid \frac{3}{4}\pi \le \varphi \le \frac{5}{4}\pi,\ \theta = \frac{\pi}{2},\ \text{or}\ \varphi \in \{\frac{3}{4}\pi, \frac{5}{4}\pi\},\ 0 \le \theta \le \frac{\pi}{2}\}$). Exact solutions are not known for the Fichera corner. Thus, we can only compute the estimated errors for the eigenfunctions and the eigenvalues.

Laplace problem

We consider the eigenvalue problem (4.8),

$$-\Delta_S u = \alpha(\alpha+1)u,$$

for the Fichera corner with pure Dirichlet boundary conditions.

An approximation for α_0 is given by $\alpha \approx 0.454168$ (computed with basis functions of polynomial degree 14). The development of the projected element and edge residuals

$$\Big\{\sum_{T\in\mathcal{T}_h} h_{+,T}^2 \|\pi_{0,T}\underline{R}_T\|_{0,T}^2\Big\}^{1/2}, \qquad \Big\{\sum_{T\in\mathcal{T}_h}\ \sum_{E\in\mathcal{E}(T)\cap\mathcal{E}_{h,\Omega_S}} \frac{\lceil E\rceil}{\lceil T\rceil} h_{+,T}^2 \|\pi_{0,E}\underline{R}_E\|_{0,E}^2\Big\}^{1/2}$$

as well as development of the error estimator $\eta = \{\sum_{T\in\mathcal{T}_h}\eta_T^2\}^{1/2}$ in the course of an adaptive refinement is displayed in Figure 4.19. The dotted line has the slope of the function $N^{-1/2}$, where N is the dimension of the discretized problem (the number of degrees of freedom, $N^{-1/2} \sim h$). The development of the approximation error $|\alpha_0 - \alpha_h|$ of the eigenvalues and the corresponding error estimator η^2 is shown in Figure 4.20, where the dotted line has the slope N^{-1} corresponding to the convergence rate $O(h^2)$. A sequence of meshes with appropriate approximate solutions u_h corresponding to the eigenvalue $\alpha_h \approx 0.454$ is displayed in Figure 4.21. Note that the mesh is refined towards the corners and the pole in each step as predicted.

Elasticity problem

We consider the eigenvalue problem (4.22) for the Fichera corner with pure Dirichlet boundary conditions. An approximation for α is given by $\alpha_0 \approx 0.522095$ (computed with basis functions of polynomial degree 10).

The element and edge residuals corresponding to the discretization of problem (4.22) are summarized in equation (4.34),

$$\underline{R}_T = \{-\nabla_S\cdot\sigma_S(\alpha_h,\underline{u}_h) + (\alpha_h-1)\,\underline{\mathbf{x}}\cdot\sigma_S(\alpha_h,\underline{u}_h)\}|_T, \quad \underline{R}_E := \lfloor\sigma_S(\alpha_h,\underline{u}_h)\cdot\mathbf{n}_E\rfloor_E\,.$$

Figure 4.25 shows the development of the residual terms

$$\Big\{\sum_{T\in\mathcal{T}_h} h_{+,T}^2 \|\pi_{0,T}\underline{R}_T\|_{0,T}^2\Big\}^{1/2}, \qquad \Big\{\sum_{T\in\mathcal{T}_h}\ \sum_{E\in\mathcal{E}(T)\cap\mathcal{E}_{h,\Omega_S}} \frac{\lceil E\rceil}{\lceil T\rceil} h_{+,T}^2 \|\pi_{0,E}\underline{R}_E\|_{0,E}^2\Big\}^{1/2}$$

and the corresponding error estimator $\eta = \{\sum_{T\in\mathcal{T}_h}\eta_T^2\}^{1/2}$ during the adaptive refinement of the mesh. The dotted line has the slope of the function $N^{-1/2}$ (convergence rate h), where $N \sim h^{-2}$ is the dimension of the discretized problem.

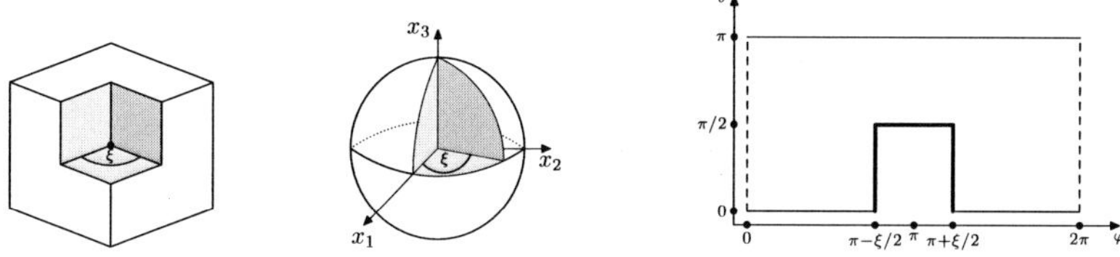

Figure 4.18: Fichera corner with angle ξ at the corner, its intersection with the unit sphere and the corresponding parameter domain (in the numerical tests: $\xi = \frac{\pi}{2}$)

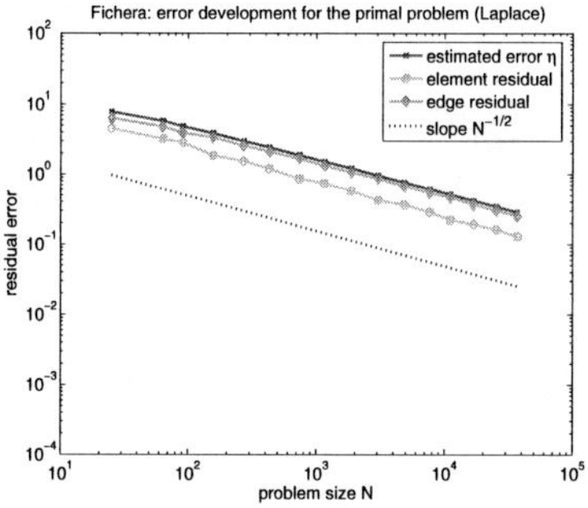

Figure 4.19: Fichera corner – development of the projected residuals and the error estimator η during an adaptive refinement for problem (4.8) for $\alpha \approx 0.454$

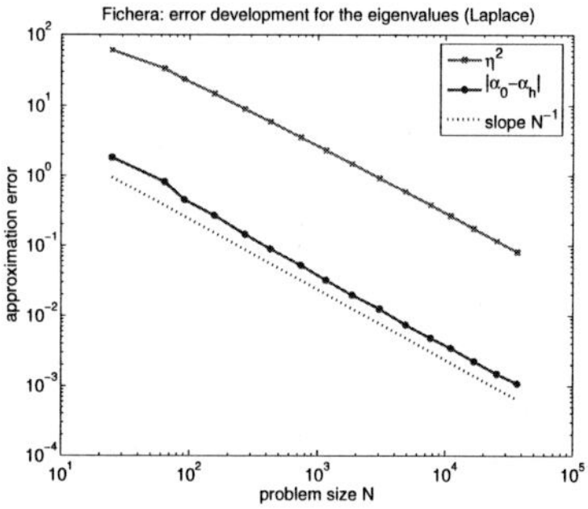

Figure 4.20: Fichera corner – development of $|\alpha_0 - \alpha_h|$ and the error estimator η^2 during an adaptive refinement for problem (4.8) for $\alpha \approx 0.454$

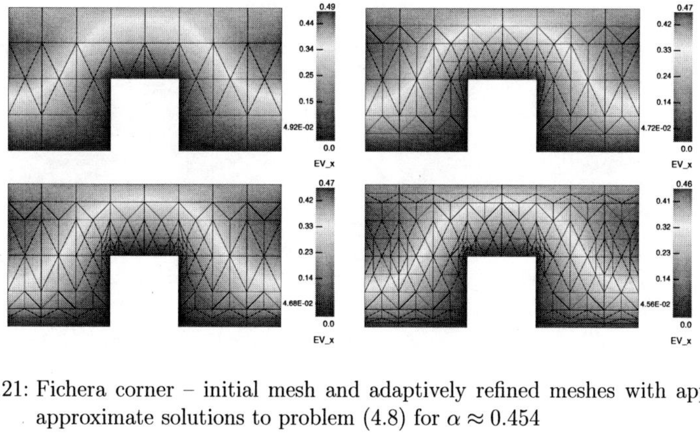

Figure 4.21: Fichera corner – initial mesh and adaptively refined meshes with appropriate approximate solutions to problem (4.8) for $\alpha \approx 0.454$

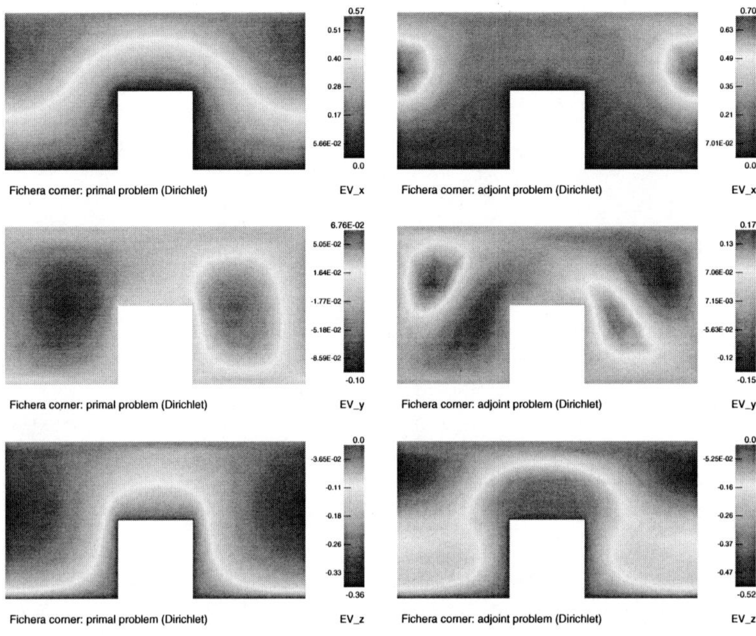

Figure 4.22: Fichera corner – components of the solutions $\underline{u}_h$ and $\underline{u}_h^\star$ to (4.22) and the corresponding adjoint eigenvalue problem for $\alpha \approx 0.522$

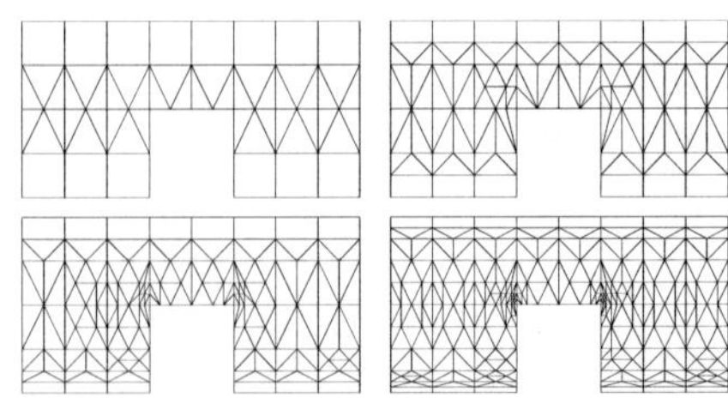

Figure 4.23: Fichera corner – initial mesh and adaptively refined meshes for problem (4.22) for $\alpha \approx 0.522$

Figure 4.24: Fichera corner – the distorted spherical domain $\underline{\mathbf{x}} + \underline{u}(\underline{\mathbf{x}})$ for problem (4.22) with $\alpha \approx 0.522$

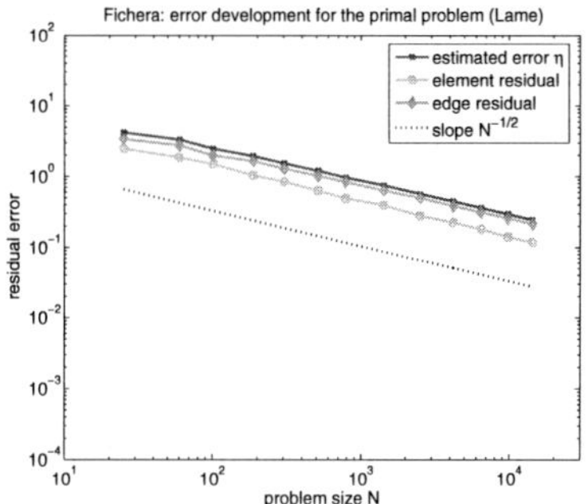

Figure 4.25: Fichera corner – development of the projected residuals and the error estimator η during an adaptive refinement for problem (4.22) for $\alpha \approx 0.522$

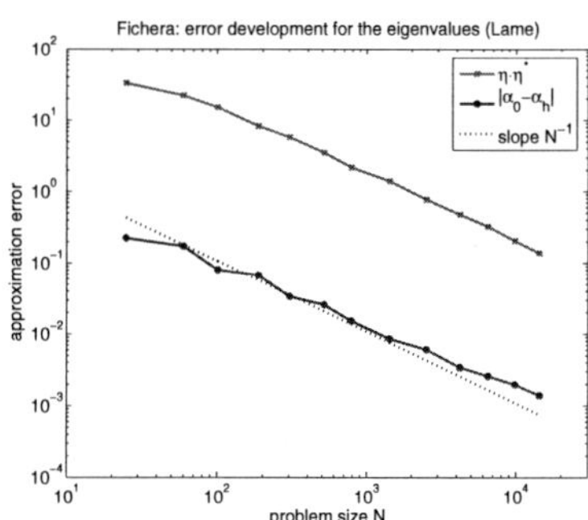

Figure 4.26: Fichera corner – development of $|\alpha_0 - \alpha_h|$ and the error estimator $\eta\,\eta^\star$ during an adaptive refinement for problem (4.22) for $\alpha \approx 0.522$

For a separate estimate for the eigenvalues of problem (4.22), the term $\eta^\star = \{\sum_{T\in\mathcal{T}_h} \eta_T^{2\,\star}\}^{1/2}$ has to be evaluated. Like for the notch (Section 4.5.1), the eigenfunction of the adjoint problem corresponding to α_0 was determined by computing the eigenfunction of problem (4.22) corresponding to $-1-\bar{\alpha}$. The development of the approximation error $|\alpha_0 - \alpha_h|$ and the corresponding error estimator $\eta\,\eta^\star$ is displayed in Figure 4.26. The dotted line has the slope of the function N^{-1} corresponding to the convergence rate $O(h^2)$.

Figure 4.22 shows the representation of the components of the approximate solutions $\underline{u}_h$ and $\underline{u}_h^\star$ to the primal and adjoint eigenvalue problems in the parameter domain corresponding to the eigenvalue $\alpha_h \approx 0.522$. The initial mesh that was used for these tests is displayed in Figure 4.23 accompanied by the meshes of the next three adaptive refinement steps. We started with a fixed number of 8 divisions into the φ-direction and 4 divisions into the θ-direction. As expected, the error is largest near the poles and the corners in the parameter domain, which is why the elements nearby are refined in each step.

In addition, the distorted spherical domain, where the computed displacements $u(\underline{\mathbf{x}})$ (of the primal problem) are added to the points $\underline{\mathbf{x}} \in \Omega_{\mathcal{S}}$, is displayed in Figure 4.24.

4.5.3 Circular cone

As a last example, we consider a circular cone with opening angle ξ to the x_3-axis as displayed in Figure 4.27 Its intersection with the unit sphere is a smooth domain without additional corners. The surface $\Omega_{\mathcal{S}}$ is parametrized by the spherical angles φ and θ. The corresponding parameter domain spanned by φ and θ is the polygon

$$\tilde{\Omega}_{\mathcal{S}} = [0, 2\pi) \times [0, \xi),$$

see Figure 4.27. The bold line in the picture corresponds to the boundary of $\Omega_{\mathcal{S}}$. The dashed lines represent periodic boundary conditions, cf. Section 3.2.

The numerical tests were done for $\xi = \frac{5}{6}\pi$ and Dirichlet boundary conditions ($\Gamma_D = \{\underline{\mathbf{x}}(\varphi,\theta) \in \partial\Omega_{\mathcal{S}} \mid \theta = \xi\}$). In general, exact solutions are not known for the circular cone, but they can be computed from a transcendental equation, see [97, 14] and references therein. Further studies regarding boundary value problems a circular cone were done, for instance, in [56, 57].

Laplace problem

We consider the eigenvalue problem (4.8), $-\Delta_S u = \alpha(\alpha+1)u$, for the circular cone with pure Dirichlet boundary conditions. An approximation for α_0 is given by $\alpha \approx 0.34618$ (computed with basis functions of polynomial degree 14).

The development of the projected element and edge residuals

$$\Big\{\sum_{T\in\mathcal{T}_h} h_{+,T}^2 \|\pi_{0,T}\underline{R}_T\|_{0,T}^2\Big\}^{1/2}, \qquad \Big\{\sum_{T\in\mathcal{T}_h}\sum_{E\in\mathcal{E}(T)\cap\mathcal{E}_{h,\Omega_S}} \frac{|E|}{|T|} h_{+,T}^2 \|\pi_{0,E}\underline{R}_E\|_{0,E}^2\Big\}^{1/2}$$

as well as development of the error estimator $\eta = \{\sum_{T\in\mathcal{T}_h} \eta_T^2\}^{1/2}$ in the course of an adaptive refinement is displayed in Figure 4.28. The dotted line has the slope of the function $N^{-1/2} \sim h$, where N is the dimension of the discretized problem. The development of the approximation error $|\alpha_0 - \alpha_h|$ of the eigenvalues and the corresponding error estimator η^2 is shown in Figure 4.29, where the dotted line has the slope N^{-1} corresponding to the convergence rate $O(h^2)$.

The initial mesh that was used for these tests is shown in Figure 4.30 accompanied by the meshes of the next three adaptive refinement steps. We started with a fixed number of 4 divisions into the φ-direction and 5 divisions into the θ-direction. In the background, the appropriate approximate solution u_h to problem (4.8) for $\alpha_h \approx 0.346$ is plotted. Since the domain Ω_S and the solution u_h are smooth, the mesh is refined almost uniformly.

Elasticity problem

We consider the eigenvalue problem (4.22) for circular cone with pure Dirichlet boundary conditions. An approximation for α is given by $\alpha_0 \approx 0.419277$ (computed with basis functions of polynomial degree 14). 4he element and edge residuals corresponding to the discretization of problem (4.22) are summarized in equation (4.34),

$$\underline{R}_T = \{-\nabla_S \cdot \sigma_S(\alpha_h, \underline{u}_h) + (\alpha_h - 1)\,\underline{\mathbf{x}} \cdot \sigma_S(\alpha_h, \underline{u}_h)\}|_T, \quad \underline{R}_E := \lfloor \sigma_S(\alpha_h, \underline{u}_h) \cdot \mathbf{n}_E \rfloor_E \,.$$

Figure 4.31 shows the development of the residual terms

$$\Big\{\sum_{T\in\mathcal{T}_h} h_{+,T}^2 \|\pi_{0,T}\underline{R}_T\|_{0,T}^2\Big\}^{1/2}, \qquad \Big\{\sum_{T\in\mathcal{T}_h}\sum_{E\in\mathcal{E}(T)\cap\mathcal{E}_{h,\Omega_S}} \frac{|E|}{|T|} h_{+,T}^2 \|\pi_{0,E}\underline{R}_E\|_{0,E}^2\Big\}^{1/2}$$

and the corresponding error estimator $\eta = \{\sum_{T\in\mathcal{T}_h} \eta_T^2\}^{1/2}$ during the adaptive refinement of the mesh. The dotted line has the slope of the function $N^{-1/2}$ (convergence rate h), where $N \sim h^{-2}$ is the dimension of the discretized problem.

For a separate estimate for the eigenvalues, the term $\eta^\star = \{\sum_{T\in\mathcal{T}_h} \eta_T^{2\,\star}\}^{1/2}$ has to be evaluated. Like for the notch (Section 4.5.1), the eigenfunction of the adjoint problem corresponding to α_0 was determined by computing an eigenfunction of problem (4.22) corresponding to $-1 - \bar{\alpha}$. The development of the approximation error $|\alpha_0 - \alpha_h|$ and the corresponding error

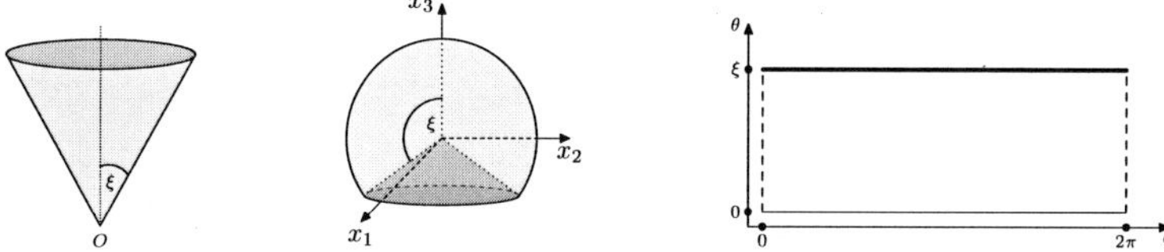

Figure 4.27: Circular cone with angle ξ to the x_3-axis (for $\xi < \frac{\pi}{2}$), the intersection with the unit sphere (for $\xi > \frac{\pi}{2}$) and the corresponding parameter domain (in the numerical tests: $\xi = \frac{5}{6}\pi$)

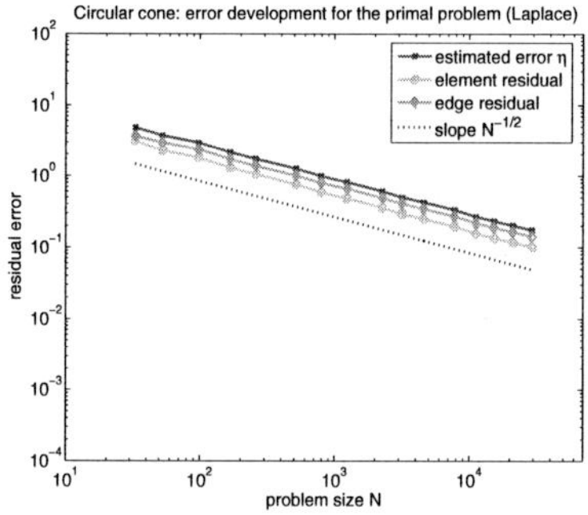

Figure 4.28: Circular cone – development of the projected residuals and the error estimator η during an adaptive refinement for problem (4.8) for $\alpha \approx 0.346$

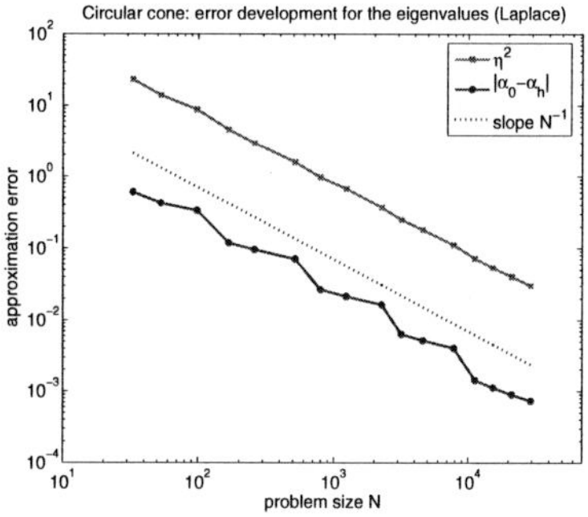

Figure 4.29: Circular cone – development of $|\alpha_0 - \alpha_h|$ and the error estimator η^2 during an adaptive refinement for problem (4.8) for $\alpha \approx 0.346$

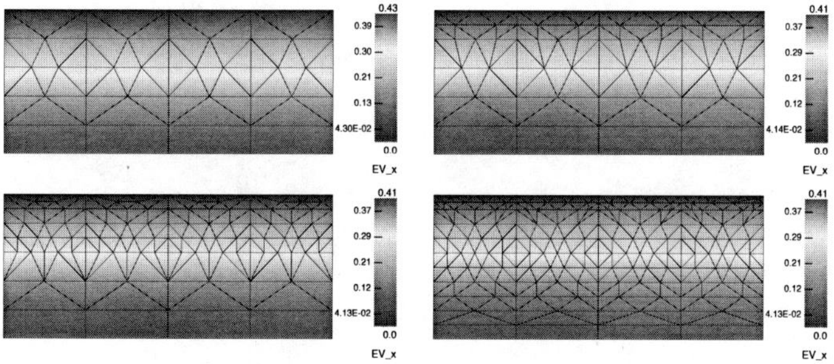

Figure 4.30: Circular cone – initial mesh and adaptively refined meshes with appropriate approximate solutions to problem (4.8) for $\alpha \approx 0.346$

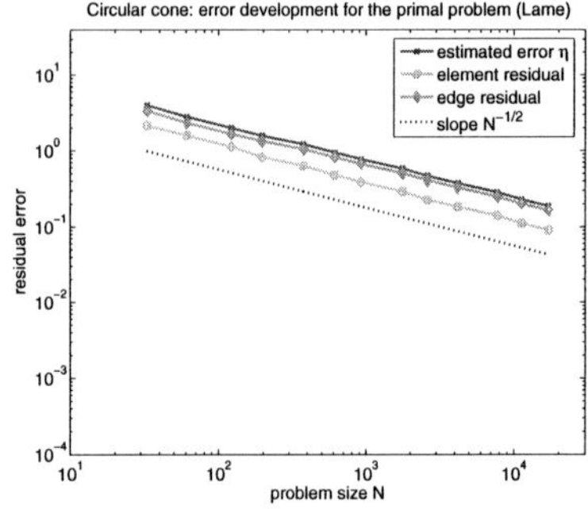

Figure 4.31: Circular cone – development of the projected residuals and the error estimator η during an adaptive refinement for problem (4.22) for $\alpha \approx 0.419$

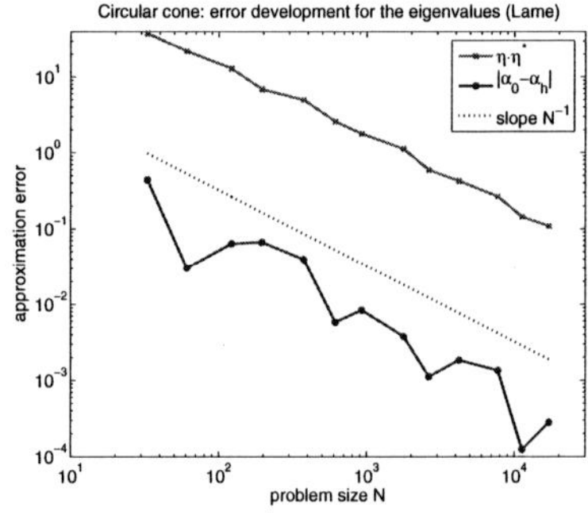

Figure 4.32: Circular cone – development of $|\alpha_0 - \alpha_h|/$ and the error estimator $\eta\,\eta^\star$ during an adaptive refinement for problem (4.22) for $\alpha \approx 0.419$

estimator $\eta\,\eta^\star$ is displayed in Figure 4.32. The dotted line has the slope of the function N^{-1} corresponding to the convergence rate $O(h^2)$.

Figure 4.33 shows the representation of the components of the approximate solutions $\underline{u}_h$ and $\underline{u}_h^\star$ to the primal and adjoint eigenvalue problems in the parameter domain corresponding to the eigenvalue $\alpha_h \approx 0.419$. The initial mesh that was used for these tests is displayed in Figure 4.34 accompanied by the meshes of the next three adaptive refinement steps. We started with a fixed number of 4 divisions into the φ-direction and 5 divisions into the θ-direction. Since the domain Ω_S and the solution $\underline{u}_h$ are smooth, the mesh is refined almost uniformly. In addition, the distorted spherical domain, where the computed displacements $u(\underline{\mathbf{x}})$ (of the primal problem) are added to the points $\underline{\mathbf{x}} \in \Omega_S$, is displayed in Figure 4.35.

Sometimes, singularity exponents greater than 1 are of interest. As a last example we consider the approximate eigenvalue $\alpha_0 \approx 1.0527$ (computed with basis functions of polynomial degree 14) for the same model problem (linear elasticity problem on a circular cone with homogeneous Dirichlet boundary, opening angle $\xi = \frac{5}{6}\pi$ (150°) to the x_3-axis). The appropriate results are printed in Figures 4.36–4.40.

4.5.4 Conclusions

We presented the results of numerous numerical examples to underline the theoretical outcomes concerning the error estimates derived in Sections 3.4, 3.5 and 3.6. As expected from a priori error estimates, the error estimator $\eta = \{\sum_{T\in\mathcal{T}_h} \eta_T^2\}^{1/2}$ decreases with the order $O(h)$, and the error estimator $\eta\,\eta^\star$ has the same convergence rate $O(h^2)$ as the eigenvalue error $|\alpha_0 - \alpha_h|$.

It is known that the optimal convergence order is obtained only when graded meshes are used. When the mesh is refined adaptively, it is sufficient to start with a uniform mesh. The error is largest near the corners of the spherical domain so that the mesh is particularly refined there anyway. This fact is perfectly justified by the numerical tests, see, for example, Figures 4.7 or 4.21. Moreover, it could be verified in further tests that no improvement

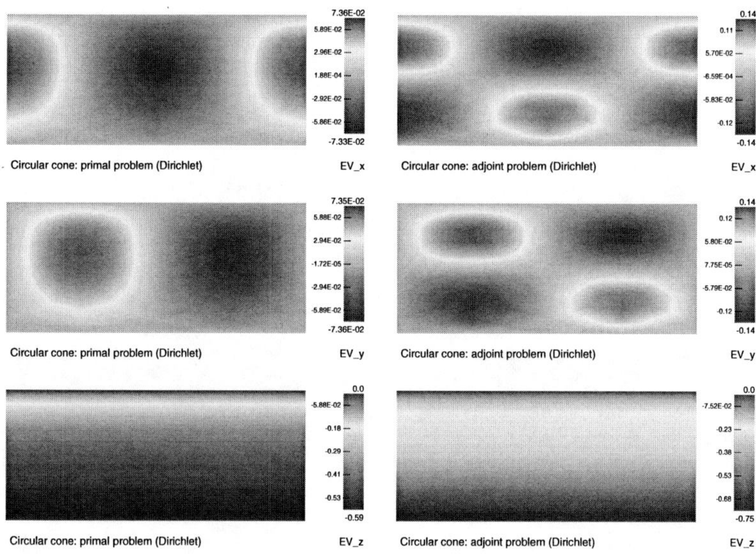

Figure 4.33: Circular cone – components of the solutions $\underline{u}_h$ and $\underline{u}_h^\star$ to (4.22) and the corresponding adjoint eigenvalue problem for $\alpha \approx 0.419$

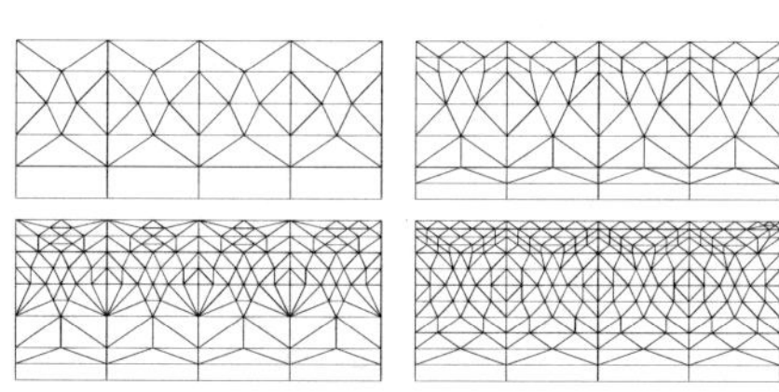

Figure 4.34: Circular cone – initial mesh and adaptively refined meshes for problem (4.22) for $\alpha \approx 0.419$

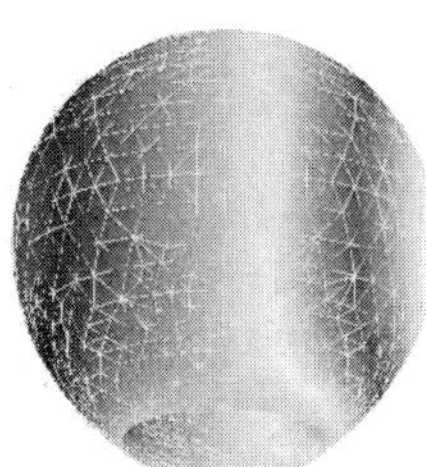

Figure 4.35: Circular cone – the distorted spherical domain $\underline{\mathbf{x}} + \underline{u}(\underline{\mathbf{x}})$ for problem (4.22) with $\alpha \approx 0.419$

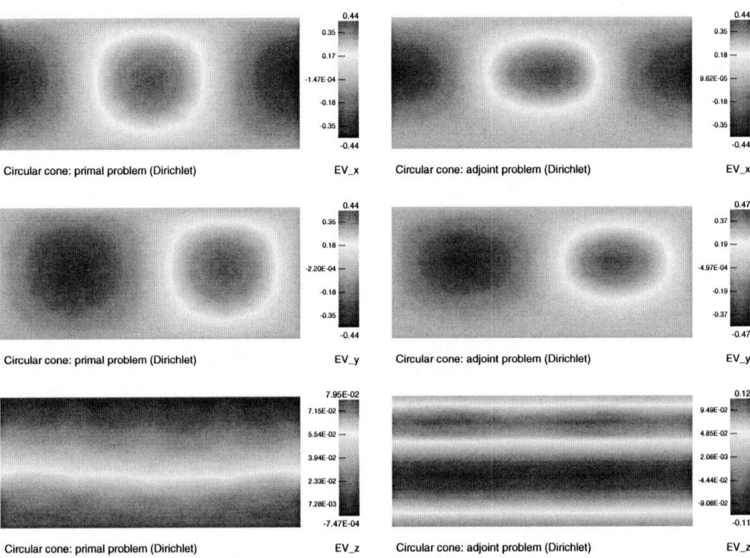

Figure 4.36: Circular cone – components of the solutions $\underline{u}_h$ and $\underline{u}_h^\star$ to (4.22) and the corresponding adjoint eigenvalue problem for $\alpha \approx 1.0527$

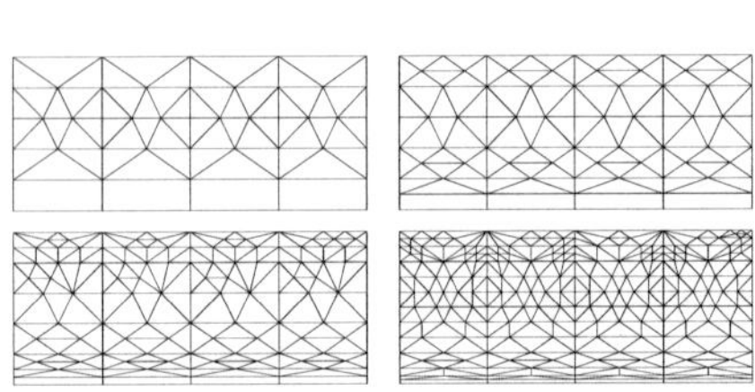

Figure 4.37: Circular cone – initial mesh and adaptively refined meshes for problem (4.22) for $\alpha \approx 1.0527$

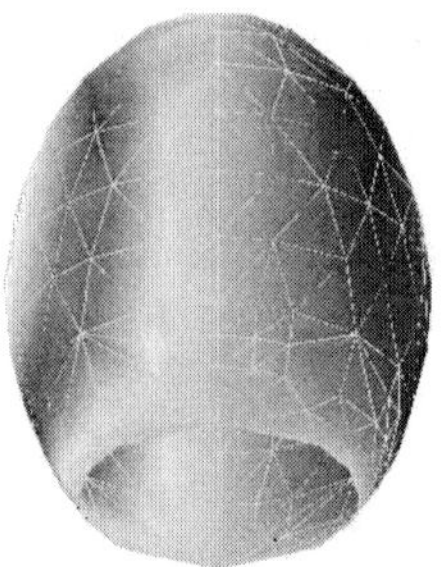

Figure 4.38: Circular cone – the distorted spherical domain $\underline{\mathbf{x}} + \underline{u}(\underline{\mathbf{x}})$ for problem (4.22) with $\alpha \approx 1.0527$

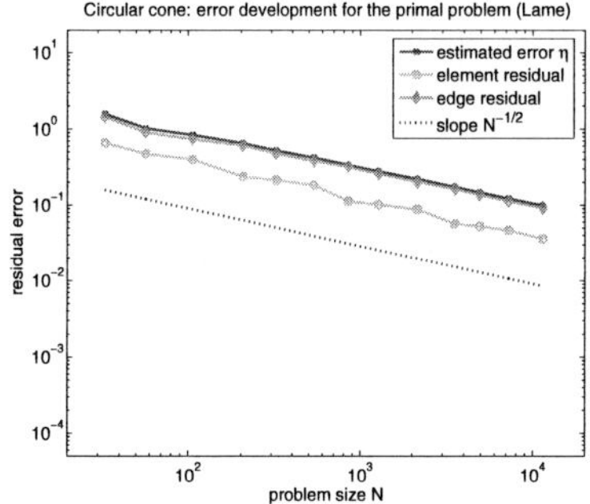

Figure 4.39: Circular cone – development of the projected residuals and the error estimator η during an adaptive refinement for problem (4.22) for $\alpha \approx 1.0527$

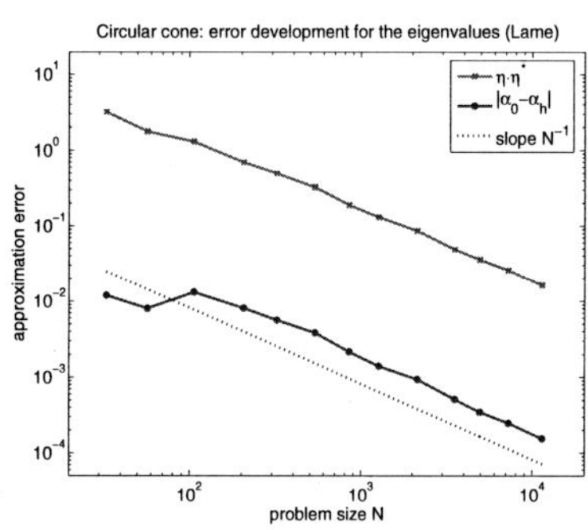

Figure 4.40: Circular cone – development of $|\alpha_0 - \alpha_h|$ and the error estimator $\eta\,\eta^\star$ during an adaptive refinement for problem (4.22) for $\alpha \approx 1.0527$

concerning the convergence behavior is achieved if graded initial meshes are used.

It was stated in Remark 3.26 that the error estimator $\tilde{\eta} = \{\sum_{T\in\mathcal{T}_h} \tilde{\eta}_T^2\}^{1/2}$ with $\tilde{\eta}$ defined in (3.13) on page 47 is reliable and efficient as well, when the element and edge residuals live in finite dimensional spaces. This is the case in the presented examples. The results of appropriate numerical tests (in particular the convergence behavior and the constants in the error estimates) differ only insignificantly from those for the projected residuals so that they are not presented on the previous pages.

Bibliography

[1] M. Ainsworth and J. T. Oden. *A Posteriori Error Estimation in Finite Element Analysis.* Wiley, New York, 2000.

[2] H. W. Alt. *Lineare Funktionalanalysis.* Springer, Berlin, Heidelberg, New York, 1999. 3. Auflage.

[3] Th. Apel, V. Mehrmann, and D. Watkins. Structured eigenvalue methods for the computation of corner singularities in 3D anisotropic elastic structures. *Comput. Methods Appl. Mech. Engrg.*, 191:4459–4473, 2002.

[4] Th. Apel, V. Mehrmann, and D. Watkins. Numerical solution of large scale structured polynomial or rational eigenvalue problems. In F. Cucker, R. DeVore, P. Olver, and E. Süli, editors, *Foundations of Computational Mathematics, Minneapolis 2002*, volume 312 of *Lecture Note Series*, Cambridge, 2004. London Mathematical Society, Cambridge University Press.

[5] Th. Apel and C. Pester. Quadratic eigenvalue problems in the analysis of cracks in brittle materials. In *Proceedings of the European Congress on Computational Methods in Applied Sciences and Engineering (ECCOMAS)*, Jyväskylä, 2004.

[6] Th. Apel and C. Pester. Clément-type interpolation on spherical domains — interpolation error estimates and application to a posteriori error estimation. *IMA J. Numer. Anal.*, 25(2):310–336, 2005.

[7] Th. Apel, A.-M. Sändig, and S. I. Solov'ev. Computation of 3D vertex singularities for linear elasticity: Error estimates for a finite element method on graded meshes. *Math. Model. Numer. Anal.*, 36:1043–1070, 2002.

[8] I. Babuska. Finite element method for domains with corners. *Computing*, 6:264–273, 1970.

[9] I. Babuška, T. Von-Petersdorff, and B. Anderson. Numerical treatment of vertex singularities and intensity factors for mixed boundary value problems for the Laplace equation in $\mathbb{R}^3$. *SIAM J. Numer. Anal.*, 31(5):1265–1288, 1994.

[10] W. Bangerth and R. Rannacher. *Adaptive Finite Element Methods for Differential Equations.* Birkhäuser, Basel, 2003.

[11] R. E. Bank. The efficient implementation of local mesh refinement algorithms. In *Adaptive computational methods for partial differential equations*, pages 74–81, College Park/Md., 1983. Proc. Workshop College Park/Md. 1983.

[12] J. R. Baumgardner and P. O. Frederickson. Icosahedral discretization of the two-sphere. *SIAM J. Numer. Anal.*, 22(6):1107–1115, 1985.

[13] Z. P. Bažant and L. F. Estenssoro. Surface singularity and crack propagation. *Int. J. Solids Structures*, 15:405–426, 1979.

[14] A. E. Beagles and A.-M. Sändig. Singularities of rotationally symmetric solutions of boundary value problems for the Lamé equations. *Z. Angew. Math. Mech.*, 71(11):423–431, 1991.

[15] P. Benner and H. Faßbender. An implicitly restarted symplectic Lanczos method for the Hamiltonian eigenvalue problem. *Lin. Alg. Appl.*, 263:75–111, 1997.

[16] P. Benner and H. Faßbender. An implicitly restarted symplectic Lanczos method for the symplectic eigenvalue problem. *SIAM J. Matrix Anal. Appl*, 22(3):682–713, 2000.

[17] J. M. Berezanskij, Z. G. Sheftel, and G. F. Us. *Functional Analysis Vol. I.* Birkhäuser, 1996.

[18] C. Bernardi. Optimal finite-element interpolation on curved domains. *SIAM J. Numer. Anal.*, 26:1212–1240, 1989.

[19] D. Braess. *Finite Elemente.* Springer, Berlin, 1997.

[20] S. C. Brenner and L. R. Scott. *The mathematical theory of finite element methods.* Springer–Verlag, New York, 1994.

[21] C. Carathéodory. *Funktionentheorie I.* Birkhäuser Verlag, 1960.

[22] C. Carstensen and S. A. Funken. Constants in Clément-interpolation error and residual based a posteriori error estimates in finite element methods. *East-West J. Numer. Math.*, 8:153–175, 2000.

[23] Ph. G. Ciarlet. *The finite element method for elliptic problems.* Oxford: North-Holland Publishing Company, Amsterdam - New York, 1978.

[24] Ph. Clément. Approximation by finite element functions using local regularization. *RAIRO Anal. Numer.*, 2:77–84, 1975.

[25] M. Costabel, M. Dauge, and Z. Yosibash. A quasidual function method for extracting edge stress intensity functions. *SIAM J. Math. Anal.*, 35(5):1177–1202, 2004.

[26] M. Cotlar and R. Cignoli. *An Introduction to Functional Analysis.* North-Holland Publishing Company, Amsterdam, 1974.

[27] J. d'Alembert. Essai d'une Nouvelle Théorie de la Résistance des Fluides. David, Paris, 1752.

[28] M. Dauge. *Elliptic boundary value problems on corner domains – smoothness and asymptotics of solutions*, volume 1341 of *Lecture Notes in Mathematics.* Springer, Berlin, 1988.

[29] R. de Boer. *Vektor- und Tensorrechnung für Ingenieure.* Springer, 1982.

[30] A. Dimitrov. On singularities in the solution of three-dimensional Stokes flow and incompressible elasticity problems with corners. *Int. J. Numer. Meth. Engrg.*, 60:773–801, 2004.

[31] A. Dimitrov, H. Andrä, and E. Schnack. Efficient computation of order and mode of corner singularities in 3D-elasticity. *Int. J. Numer. Methods Engrg.*, 52:805–827, 2001.

[32] A. Dimitrov, H. Andrä, and E. Schnack. Singularities near three-dimensional corners in composite laminates. *Int. J. Fracture*, 115(4):361–375, 2002.

[33] A. Dimitrov and E. Schnack. Asymptotical expansion in non-Lipschitzian domains: a numerical approach using h-fem. *Numer. Linear Algebra Appl.*, 00:1–18, 2002.

[34] A. Dimitrov and E. Schnack. Singularities of surface-breaking cracks in bi-material interfaces. In *W. Wendland (ed.) et al., Analysis and simulation of multifield problems. Selected papers of the international conference on multifield problems, Stuttgart, Germany, April 8-10, 2002, Lect. Notes Appl. Comput. Mech. 12, 199-204*. Berlin: Springer, 2003.

[35] T. Dupont and R. Scott. Polynomial approximation of functions in Sobolev spaces. *Math. Comp.*, 34:441–463, 1980.

[36] G. Dziuk. Finite elements for the Beltrami operator on arbitrary surfaces. In *Partial differential equations and calculus of variations*, volume 1357 of *Lecture Notes in Math.*, pages 142–155, Berlin, 1988. Springer.

[37] L. Euler. Principes généraux du mouvement des fluides. *Mémoires de l'académie des sciences de Berlin*, 11:274–315, 1757. English summary in Opera Omnia (E226): Series 2, Volume 12, pp. 54–91.

[38] G. Fichera. Asymptotic behaviour of the electric field and density of the electric charge in the neighbourhood of singular points of a conducting surface. *Russian Math. Surveys*, 30(3):107–127, 1975.

[39] R. W. Freund. Lanczos-type algorithms for structured non-Hermitian eigenvalue problems. In J. D. Brown, M. T. Chu, D. C. Ellison, and R. J. Plemmons, editors, *Proceedings of the Cornelius Lanczos International Centenary Conference*, pages 243–245. SIAM, 1994.

[40] K. Giebermann. *Schnelle Summationsverfahren zur numerischen Lösung von Integralgleichungen für Streuprobleme im* $\mathbb{R}^3$. PhD thesis, Universität Karlsruhe, Germany, 1997.

[41] P. Grisvard. Behavior of the solutions of an elliptic boundary value problem in a polygonal or polyhedral domain. In B. Hubbard, editor, *Proc. 3rd Symp. Numer. Solut. Partial Differ. Equat., 1975*, pages 207–274. Academic Press, New York, 1976.

[42] P. Grisvard. *Elliptic problems in nonsmooth domains.* Monographs and Studies in Mathematics, 24. Pitman Advanced Publishing Program. Boston-London-Melbourne: Pitman Publishing, 1985.

[43] S. Großmann. *Funktionalanalysis.* Aula-Verlag, Wiesbaden, 1988.

[44] V. Heuveline and R. Rannacher. A posteriori error control for finite element approximations of elliptic eigenvalue problems. *Journal on Advances in Computational Mathematics. Special issue "A posteriori Error Estimation and Adaptive Computational Methods"*, 15(1–4), 2001.

[45] T. Hwang, W. Lin, and V. Mehrmann. Numerical solution of quadratic eigenvalue problems with structure-preserving methods. *SIAM J. Sci. Comp.*, 24:1283–1302, 2003.

[46] L. Jentsch. On the elastic potentials at corners. *Asymptotic Analysis*, 14:73–95, 1997.

[47] M. Kachanov, B. Shafiro, and I. Tsukrov. *Handbook of elasticity solutions.* Kluwer, 2003.

[48] O. O. Karma. Approximation in eigenvalue problems for holomorphic Fredholm operator functions. I. *Numer. Funct. Anal. Optimization*, 17:365–387, 1996.

[49] O. O. Karma. Approximation in eigenvalue problems for holomorphic Fredholm operator functions. II: Convergence rate. *Numer. Funct. Anal. Optimization*, 17:389–408, 1996.

[50] W. T. Koiter. On the foundations of the linear theory of thin elastic shells. I. *Proc. Kon. Ned. Akad. Wetensch.*, 1970.

[51] V. A. Kondrat'ev. Boundary problems for elliptic equations in domains with conical or angular points. *Trans. Moscow Math. Soc.*, 16:227–313, 1967. English translation.

[52] M. M. Konstantinov, V. Mehrmann, and P. Hr. Petkov. Perturbation analysis of Hamiltonian Schur and Block-Schur forms. *SIAM J. Matrix Anal. Appl.*, 23:387–424, 2001.

[53] V. A. Kozlov and V. G. Maz'ya. *Differential equations with operator coefficients with applications to boundary value problems for partial differential equations.* Springer, 1999.

[54] V. A. Kozlov, V. G. Maz'ya, and J. Roßmann. Spectral properties of operator pencils generated by elliptic boundary value problems for the Lamé system. *Rostocker Math. Kolloq.*, 51:5–24, 1997.

[55] V. A. Kozlov, V. G. Maz'ya, and J. Roßmann. *Spectral problems associated with corner singularities of solutions to elliptic equations.* Mathematical Surveys and Monographs, Volume 85, American Mathematical Society, 2001.

[56] V. A. Kozlov, V. G. Maz'ya, and C. Schwab. On singularities of solutions of the displacement problem of linear elasticity near the vertex of a cone. *Arch. Ration. Mech. Anal.*, 119(3):197–227, 1992.

[57] V. A. Kozlov, V. G. Maz'ya, and C. Schwab. On singularities of solutions to the Dirichlet problem of hydrodynamics near the vertex of a cone. *J. Reine Angew. Math.*, 456:65–97, 1994.

[58] D. Kressner. Perturbation bounds for isotropic invariant subspaces of skew-Hamiltonian matrices. *SIAM J. Matrix Anal. Appl*, 26(4):947–961, 2005.

[59] A. Kufner and A.-M. Sändig. *Some Applications of Weighted Sobolev Spaces.* Teubner, Leipzig, 1987.

[60] G. Kunert. *A posteriori error estimation for anisotropic tetrahedral and triangular finite element meshes.* PhD thesis, TU Chemnitz, Logos, Berlin, 1998.

[61] S. Lang. *Complex Analysis.* Addison Wesley Publishing Company, 1977.

[62] M. G. Larson. A posteriori and a priori error analysis for finite element approximations of self-adjoint elliptic eigenvalue problems. *SIAM J. Numer. Anal.*, 38(2):608–625, 2000.

[63] A. T. Layton. Cubic spline collocation method for the shallow water equations on the sphere. *J. Comput. Phys.*, 179(2):578–592, 2002.

[64] D. Leguillon. Computation of 3d singularities in elasticity. In M. Costabel, M. Dauge, and S. Nicaise, editors, *Boundary value problems and integral equations in nonsmooth domains*, volume 167 of *Lecture notes in pure and applied mathematics*, pages 161–170, New York, 1995. Marcel Dekker. Proceedings of a conference at CIRM, Luminy, France, May 3-7, 1993.

[65] H. Leipholz. *Einführung in die Elastizitätstheorie.* G. Braun, Karlsruhe, 1968.

[66] J. L. Lions and E. Magenes. *Non-homogeneous boundary value problems and applications. Vol. I.* Die Grundlehren der mathematischen Wissenschaften. Band 181. Berlin-Heidelberg-New York: Springer-Verlag, 1972.

[67] A. E. H. Love. *A treatise on the mathematical theory of elasticity. 4th ed.* New York: Dover Publications. XVIII, 643 p., 1944.

[68] L. E. Malvern. *Introduction to the Mechanics of a Continuous Medium.* Prentice Hall, Inc. Englewood Cliff, New Jersey, 1969.

[69] A. S. Markus. *Introduction to the spectral theory of polynomial operator pencils. Transl. from the Russian by H. H. McFaden.* Translations of Mathematical Monographs, 71. Providence, RI: American Mathematical Society (AMS), 1988.

[70] V. G. Maz'ya and B. A. Plamenevskiĭ. The first boundary value problem for classical equations of mathematical physics in domains with piecewise smooth boundaries, part I, II. *Z. Anal. Anwend.*, 2:335–359, 523–551, 1983. In Russian.

[71] W. McLean. *Strongly Elliptic Systems and Boundary Integral Equations.* Cambridge University Press, 2000.

[72] V. Mehrmann and D. Watkins. Structure-preserving methods for computing eigenpairs of large sparse skew-Hamiltonian/Hamiltonian pencils. *SIAM J. Sci. Comp.*, 22(6):1905–1925, 2001.

[73] V. Mehrmann and D. Watkins. Polynomial eigenvalue problems with Hamiltonian structure. *Electron. Trans. Numer. Anal.*, 13:106–118, 2002.

[74] A. Meyer. Personal communication, 2002–2005.

[75] A. Meyer and C. Pester. The Laplace and the linear elasticity problems near polyhedral corners and associated eigenvalue problems. Preprint SFB393/04-12, Preprint-Reihe des SFB393 der Technischen Universität Chemnitz, 2004.

[76] J. Mu. Solving the Laplace-Beltrami equation on S^2 using spherical triangles. *Numer. Methods Partial Differential Equations*, 12(5):627–641, 1996.

[77] N. I. Mußchelischwili. *Einige Grundaufgaben zur mathematischen Elastizitätstheorie.* VEB Fachbuchverlag Leipzig, 1971.

[78] P. M. Naghdi. Foundations of elastic shell theory. *Progress in Solid Mechanics*, 4:1–90, 1963.

[79] S. A. Nazarov and B. A. Plamenevsky. *Elliptic problems in domains with piecewise smooth boundary*, volume 13 of *de Gruyter Expositions in Mathematics.* Walter de Gruyter, Berlin, 1994.

[80] P. Neittaanmäki and S. Repin. *Reliable Methods for Computer Simulation: Error Control and A Posteriori Estimates.* Elsevier, 2004.

[81] S. Nemat-Nasser and M. Hori. *Micromechanics: Overall Properties of Heterogeneous Solids.* Amsterdam Elsevier, 1993.

[82] S. Nicaise and A.-M. Sändig. Transmission problems for the Laplace and elasticity operators: Regularity and boundary integral formulation. *Math. Models Methods Appl. Sci.*, 9(6):855–898, 1999.

[83] N. Omer, Z. Yosibash, M. Costabel, and M. Dauge. Edge flux intensity functions in polyhedral domains and their extraction by a quasidual function method. *Int. J. Fracture*, 129:97–130, 2004.

[84] J. P. Papadakis and I. Babuska. A numerical procedure for the determination of certain quantities related to the stress intensity factors in two-dimensional elasticity. *Comput. Methods Appl. Mech. Engrg.*, 122(1–2):69–92, 1995.

[85] H. Parisch. *Festkörper-Kontinuumsmechanik: Von den Grundgleichungen zur Lösung mit Finiten Elementen.* Teubner, Stuttgart, 2003.

[86] E. Peschl. *Differentialgeometrie.* Bibliographisches Institut, Mannheim, 1973.

[87] C. Pester. CoCoS – Computation of corner singularities. Preprint SFB393/05-03, Preprint-Reihe des SFB393 der Technischen Universität Chemnitz, 2005. Dokumentation.

[88] C. Pester. Hamiltonian eigenvalue symmetry for quadratic operator eigenvalue problems. *J. Integral Equations Appl.*, 17(1), 2005.

[89] M. Pester. Parallelrechner-Bibliotheken. Internet publication, `http://www-user.tu-chemnitz.de/~pester/par_lib.html`, 1996.

[90] S. Prößdorf. *Einige Klassen singulärer Gleichungen.* Akademie-Verlag, Berlin, 1974.

[91] A. Quarteroni and A. Valli. *Numerical approximation of partial differential equations.* Springer, Berlin, 1994.

[92] M. Reed and B. Simon. *Methods of modern mathematical physics. I: Functional analysis.* Academic Press, New York etc., 1980.

[93] F. Riesz. Über lineare Funktionalgleichungen. *Acta Mathematica*, 41:71–98, 1918.

[94] J. Rosam. Berechnung der Rissgeometrie bei spröden elastischen Körpern. Diplomarbeit, TU Chemnitz, 2004.

[95] W. Rudin. *Real and complex analysis. 2nd ed.* McGraw-Hill Series in Higher Mathematics. McGraw-Hill Book Comp., 1974.

[96] A. F. Ruston. *Fredholm theory in Banach spaces.* Cambridge University Press, 1986.

[97] A.-M. Sändig, U. Richter, and R. Sändig. The regularity of boundary value problems for the Lamé equations in a polygonal domain. *Rostock. Math. Kolloq.*, 36:21–50, 1989.

[98] A.-M. Sändig and R. Sändig. Singularities of non-rotationally symmetric solutions of boundary value problems for the Lamé equations in a 3 dimensional domain with conical points. In B.-W. Schulze, editor, *Symposium "Analysis on manifolds with singularities", Breitenbrunn 1990.*, volume 131, pages 181–193. Teubner, 1992.

[99] S. Sauter and Chr. Schwab. *Randelementmethoden. Analyse, Numerik und Implementierung schneller Algorithmen.* Teubner, 2004.

[100] H. Schmitz, K. Volk, and W. L. Wendland. On three-dimensional singularities of elastic fields near vertices. *Numer. Methods Partial Differential Equations*, 9:323–337, 1993.

[101] W. Schöne. *Differentialgeometrie.* Teubner, Leipzig, 1987.

[102] L. K. Steger. *Sphärische finite Elemente und ihre Anwendung auf Eigenwertprobleme des Laplace-Beltrami-Operators.* PhD thesis, Ludwig-Maximilian-Universität München, 1983.

[103] F. Tisseur. Stability of structured Hamiltonian eigensolvers. *SIAM J. Matrix Anal. Appl*, 23(1):103–125, 2001.

[104] R. Verfürth. *A review of a posteriori error estimation and adaptive mesh-refinement techniques.* Wiley and Teubner, Chichester and Stuttgart, 1996.

[105] D. Watkins. *Fundamentals of Matrix Computations.* John Wiley & Sons, 2002.

[106] D. Watkins. On Hamiltonian and symplectic Lanczos processes. *Lin. Alg. Appl.*, 385:23–45, 2004.

[107] J. Weidmann. *Lineare Operatoren in Hilbertäumen. Teil I: Grundlagen.* Teubner, 2000.

[108] W. L. Wendland. Strongly elliptic boundary integral equations. In *The state of the art in numerical analysis, Proc. Joint IMA/SIAM Conf., Birmingham/Engl. 1986, Inst. Math. Appl. Conf. Ser., New Ser. 9, 511–562*, 1987.

[109] D. Werner. *Funktionalanalysis.* Springer, Berlin, Heidelberg, New York, 5th, expanded edition, 2005.

[110] Wikipedia. Cauchy-Riemann equations. Internet source. `http://en.wikipedia.org/wiki/Cauchy_Riemann_equations`.

[111] Z. Yosibash, R. Actis, and B. Szabó. Extracting edge flux intensity functions for the Laplacian. *Int. J. Numer. Meth. Engrg.*, 53(1):225–242, 2002.

[112] K. Yosida. *Functional analysis.* Die Grundlehren der mathematischen Wissenschaften in Einzeldarstellungen. 123. 3rd ed. Springer, 1971.

[113] E. Zeidler. *Applied functional analysis. Applications to mathematical physics. Vol. 1.* Applied Mathematical Sciences. 108. Springer, 1995.

Index

Theses

A posteriori error estimation for non-linear eigenvalue problems for differential operators of second order with focus on $3D$ vertex singularities

Dipl. Math. Cornelia Pester

Institut für Mathematik und Bauinformatik
Fakultät für Bauingenieur- und Vermessungswesen
Universität der Bundeswehr München

1. The solution to an elliptic boundary value problem in a domain Ω with polyhedral corners has not the typical Sobolev regularity as known for problem in smooth domains. It is composed of a regular part U_r and a singular part U_s. If the differential operator defining the boundary value problem is of second order, then typically $U_r \in H^2(\Omega)$ and $U_s \notin H^2(\Omega)$. Thus, U_s is the solution to the boundary value problem with a homogeneous right-hand side. Corner singularities are of local nature. For the description of U_s, it can be assumed, without loss of generality, that Ω coincides with an infinite cone with one corner.

2. The singular part U_s can be written as a series with terms of the form $r^{\alpha_i} u_i$ (and further logarithmic terms), where r is the distance to the corner and α_i are the singularity exponents. The exponents α_i and the functions u_i are eigenpairs of a polynomial eigenvalue problem which can be derived from the given boundary value problem. There might be complex eigenpairs $[\alpha_i, u_i]$ although the original problem possesses only real solutions.

3. The eigenvalue problem for α and u is defined on the surface of a ball which is centered at the corner of the conical domain. For the analysis of this eigenvalue problem, it is necessary to parametrize the (unit) sphere. Spherical coordinates have many convenient features and are therefore used in this thesis; but other parametrizations are conceivable as well. One advantage of spherical coordinates is that they generate a global parameter domain $\tilde{\Omega}_{\mathcal{S}}$ in $\mathbb{R}^2$ corresponding to the spherical domain $\Omega_{\mathcal{S}} = \Omega \cap \mathcal{S}^2$ so that two-dimensional considerations are applicable in most cases.

4. Normally, the exact solutions to the mentioned eigenvalue problem cannot be computed with analytic means. Numerical methods have to be employed to obtain an approximate solution. The finite element method is a useful tool to discretize and to solve such an eigenvalue problem. One is interested in the error between the exact solutions and the approximate solutions. As the exact solutions are not known, this error has to be estimated. In this thesis, appropriate a posteriori error estimates are derived for eigenvalue problems which are defined on the unit sphere. The presented theory includes but is not limited to eigenvalue problems which stem from elliptic boundary

value problems in a cone. In general, eigenvalue problems for real- or complex-valued vector functions are studied, which can be written as operator eigenvalue problems for holomorphic Fredholm operator pencils.

The consideration of complex-valued functions causes restrictions in comparison with the real case, especially concerning the differentiation. A complex-valued function is differentiable if and only if it satisfies the Cauchy-Riemann equations. Functions lie $\bar{z}$ or $|z|^2$ are not differentiable. To discard the trivial solution to the eigenvalue problem, one usually requires $\|u\| = 1$ for the eigenfunctions u and some norm $\|\cdot\|$. For the verification of the upper error bound, the term $\|u\|^2$ has to be derived with respect to u. Independently of the specific choice of the norm, the derivative does not exist in a complex Hilbert space. A remedy can be found in requiring $(u, \bar{u}) = 1$ for some inner product $(\cdot, \cdot)$. The specific structure of this inner product is irrelevant, but potential solutions $u \neq 0$ with $(u, \bar{u}) = 0$ are skipped in this way and cannot be treated within the framework of this thesis.

5. For the discretization of the eigenvalue problem, a triangulation of the spherical domain $\Omega_\mathcal{S} \subset \mathcal{S}^2$ has to be defined. Concerning the implementation and the comparability to the two-dimensional case, it is useful to define the triangulation of $\Omega_\mathcal{S}$ so that the parameter domain $\tilde{\Omega}_\mathcal{S}$, which is spanned by the spherical angles $\varphi \in [0, 2\pi)$ and $\theta \in [0, \pi]$, is divided into elements (usually triangles) with straight-lined boundary. Then, a strategy similar to the route for the planar case allows us to derive a *reliable* and *efficient* error estimator for the eigenpairs.

   In order to avoid a reliability–efficiency gap between the upper and lower error bounds, the spherical domain $\Omega_\mathcal{S}$ has to be divided into *isotropic* elements, this means into elements which have approximately the same spatial dimensions in all directions. Then, the corresponding triangulation of the parameter domain consists of anisotropic elements. Elements at the poles of the sphere ($\theta = 0$ or $\theta = \pi$) are identified with rectangles in the parameter domain.

6. For the definition of a finite element space, nodal basis functions are used. In accordance with the planar case, the nodal basis functions can be defined over the elements in the parameter domain so that they are affine linear with respect to the spherical angles φ and θ. For pole elements, affine bilinear nodal basis functions are needed.

7. The nodal basis functions are used to define a Clément-type interpolation operator. Like all norms and operators which are defined on the sphere, the interpolation operator is weighted with a certain factor resulting from the spherical surface element. For spherical coordinates, this weight is the term $\sin\theta$ which can conveniently be approximated by θ or $\pi - \theta$ if necessary. This allows for a simplification of the verification of certain estimates. In this manner, a trace theorem and a Poincaré-type inequality on the sphere can be proven. From these two auxiliary results, local interpolation error estimates for the Clément-type operator can be concluded.

   The obtained estimates are similar to those known for planar domains and are therefore not surprising. Nevertheless, many details have to be proven anew because the spherical nature of the problem causes several difficulties which require special care.

8. The interpolation error estimates are used to derive a residual a posteriori error estimator which provides an upper bound for the error between the eigenpairs of the original

and the discretized eigenvalue problems. If the residuals live in a finite-dimensional space, then the computed error estimator provides a lower error bound as well. Otherwise, the residuals have to be projected into a finite-dimensional space in order to prove the efficiency of the error estimator. Then, lower and upper error bounds are obtained up to the approximation errors between the residuals and their projections. These approximation errors, however, are of higher order and do not influence the numerical results essentially.

9. The computed error estimator gives an approximation of the error $|\lambda_0-\lambda_h|+\|u_0-u_h\|_V$, where $[\lambda_0, u_0]$ is an exact eigenpair and $[\lambda_h, u_h]$ is a solution to the discretized eigenvalue problem and an approximation to $[\lambda_0, u_0]$.

   From a priori error estimates, it is known that the eigenvalues converge faster than the eigenfunctions. With eigenfunctions and residuals from the dual problem, the error estimate for the eigenvalues can be improved.

10. For the derivation and verification of the error estimator, a nonlinear operator function $F : \mathbb{K} \times V \longrightarrow \mathbb{K} \times V^\star$ is introduced whose Fréchet derivative influences the constants in the error estimates essentially. When $\lambda_0 \in \mathbb{K}$ is an eigenvalue of the original eigenvalue problem with the corresponding eigenfunction u_0, then the Fréchet derivative $\mathrm{D}F([\lambda_0, u_0])$ of F at $[\lambda_0, u_0]$ is a linear homeomorphism (invertible so that the inverse operator is continuous) *if and only if* λ_0 is a simple eigenvalue.

    The bounded inverse of the Fréchet derivative is needed only to prove the reliability of the computed error estimator. If λ_0 is not a simple eigenvalue, nothing can be said about the validity of the upper error bound. Numerical experiments show that reasonable results can be obtained also for multiple eigenvalues, which gives rise to assume that the constants in the error estimates are not optimal. Nevertheless, it is possible to outline the terms on which the constants in each of the derived estimates depend.

11. The obtained error estimates are explained considering two examples, the Laplace and the linear elasticity problems in conical domains. In each case, we derive the associated eigenvalue problem for the singularity exponents α and the functions u from the approach $U = r^\alpha u$. While the eigenvalue problem corresponding to the Laplace equation can be reduced to a linear eigenvalue problem, the eigenvalue problem derived from the Lamé equations is quadratic in α.

12. It is known from a priori error estimates that the optimal convergence order of the eigenvalues and eigenfunctions is obtained only for graded finite element meshes. When the mesh is refined adaptively on the basis of the computed error estimator, graded meshes can be abandoned, because the mesh is refined automatically towards the points in question. Numerous numerical experiments concerning the two eigenvalue problems mentioned in thesis 11 were carried out. The obtained results confirm the theoretical outcomes.

13. The presented theory can be extended to a posteriori error estimation for problems on arbitrary two-dimensional surfaces provided that certain assumptions on the triangulation (in particular the isotropy) are satisfied and that auxiliary results like a Poincaré-type inequality and the trace theorem hold.